Preface

A fresh approach to company law

A company is merely a device created by human minds to achieve identified goals: nothing can be understood unless that central fact is recognised. This is a book in which, to use EM Forster's idea, I seek to connect the prose with the passion in company law. Or to put the same idea more simply, I am trying to bring the technical rules to life. It is a book that is intended primarily to be used by university students seeking an overview of the subject before class, or seeking a guide through revision, or seeking some context within which to understand company law more generally to provide them with pathways into further reading. Above all else this is a critical essay about the nature of company law in the modern world. Nevertheless, there is sufficient technical detail to satisfy professionals seeking either a refresher or an introduction to the subject, or students wanting assistance with particular company law problems. Yet there is also enough theoretical analysis to present a single coherent view of company law at the beginning of the 21st century.

This book is accompanied, as are all of my books, by a series of podcasts that parallel an entire undergraduate course in company law on my website at: www.alastairhudson.com; just follow the links to the 'company law' area. There are also vidcasts hosted on that site, as well as advice on answering exam questions prepared originally for my students in the University of London, and podcasts and essays that probe into further issues.

The scope of this book

Traditionally company law considers the legal nature of companies, the principal legislation governing the operation of companies, the regulations governing the ways in which companies raise capital, the rights of shareholders, and the duties of company directors. This book, however, extends beyond that traditional material to consider the role of corporate governance standards and the rapidly growing topic of corporate social responsibility.

We begin with a discussion of the history of the company and the development of the *Salomon* principle of separate legal personality. Aron Salomon was a bootmaker in London's East End in the 19th century, and the litigation about the debts incurred by his business is the foundation stone for modern company law. In homage to that great case, there is a pair of boots on the cover of this book. That case established the principle that a company is a distinct legal entity: something that has profound legal, commercial and moral ramifications that continue to be of huge significance in our global economy today. This principle is examined in Chapters 2 and 3.

Then the book explains how a company is structured, set up and run (Chapters 4 to 6). All the while, however, we shall be considering how companies are used in the modern world, and how company law theory tends to be locked into a narrow conception of the company. We will then spend a large proportion of our time considering the legal obligations of directors (Chapter 7) and the rights of shareholders (Chapter 8) – both of which are fascinating in terms of the differences between the decided cases, in terms of the recent development of the statutory principles and simply at the conceptual level.

In essence, I think that the interaction of directors and shareholders through the prism of 'the company as a distinct legal person' is the beating heart of company law. Consequently it forms a large part of this book. Related to those issues is the new vogue for studying corporate governance, which used to be considered to be outside company law but which has increased in importance in practice. Once we have considered the human dramas within companies between directors and shareholders, we turn to the organisation of the company's capital and its shares.

From there it is a short hop to the growing field of securities law, which governs the means by which large public companies raise capital so as to fund their activities. This is a very new, parallel dimension to mainstream company law and one which most of the company law books have yet to embrace fully. The securities field is part of a new zone in which companies are regulated by statutory and other bodies, which is considered in Chapter 11. More high finance is considered in Chapter 12 in relation to mergers and takeovers. When these issues of finance law are grafted onto traditional company law, we have to recognise that a new beast emerges. Having moved through the birth and life of companies, we come to the death of companies by means of corporate insolvency (Chapter 13) and the means by which outsiders recover (or do not recover) what they are owed.

To that extent, the book may seem to have moved from a contextual, historical discussion to a large amount of discussion of technical matters. However, these later chapters consider the company's place in the larger world, and its economic, political and ethical place in society more generally. The final chapters of the book consider corporate social responsibility head-on. This growing field (which is important now in the management and/or the public relations of large companies, depending on your view) relates particularly (but not exclusively) to the activities of larger companies and their effects on developing economies, their workforces and the environment. The final chapter then draws together much of the critique of company law that runs through this book like a seam of iron ore

Contents

Understanding Company Law

Alastair Hudson
LLB, LLM, PhD, FRSA, FHEA

Professor of Equity & Law
Queen Mary University of London
Barrister, Lincoln's Inn
National Teaching Fellow

Routledge
Taylor & Francis Group
LONDON AND NEW YORK

First published 2012
by Routledge
2 Park Square, Milton Park, Abingdon, Oxon OX14 4RN

Simultaneously published in the USA and Canada
by Routledge
711 Third Avenue, New York, NY 10017

Routledge is an imprint of the Taylor & Francis Group, an informa business

British Library Cataloguing in Publication Data
A catalogue record for this book is available
from the British Library

Library of Congress Cataloging in Publication Data
Hudson, Alastair.
 Understanding company law / Alastair Hudson.
  p. cm.
 Includes bibliographical references and index.
 1. Corporation law – Great Britain.
I. Title.
 KD2079.H83 2011
 346.41'066–dc22

2011009515

ISBN: 978–0–415–68217–6 (hbk)
ISBN: 978–0–415–68218–3 (pbk)
ISBN: 978–0–203–14830–3 (ebk)

Typeset in Times
by RefineCatch Limited, Bungay, Suffolk

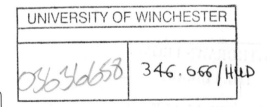

Printed and bound in Great Britain by the MPG Books Group

through rock. To think properly about company law is to think about the whole of the modern world. It is a fascinating zone of engagement.

The writing of this book

I have written this book for anyone who is interested in company law. It covers all of the key ideas that one might need to confront for a university course in company law. It does not pretend to be a full textbook, although most of the major cases, schools of thought and statutory provisions are considered here. I am a teacher and so what I will attempt to do is to teach you the key principles of company law in this book. All good teaching, however, must also seek to enthuse the learner, to open her mind, and to point out interesting paths through the thicket of ideas. So, what this book also tries to do is to encourage those with an interest in the subject to think about what company law is, how it relates to the rest of our law, and how it affects our world. If I can communicate a little of my enthusiasm for the subject to you, then I have achieved one of the key goals of this book. In my own mind, this book was intended to be a long, elegant essay on modern company law, and it has been a joy to write. In it, I have been able to meet lots of favourite ideas as though they were old friends.

This book was written at various times during 2009 and 2010, frequently late at night after the rigours of the day were over, before being finished in January 2011. The law is as I found it from the materials available to me during that time.

I would like to offer thanks to a number of people, in no particular order. To Fiona Kinnear of Routledge who listened to rumours of this book for some time, and who is a great publisher. To those who first lit a fire in me for this subject as a student: Harry Rajak and Eva Lomnicka. They bear no blame, however, for what has emerged between these covers. To colleagues whose scholarship and camaraderie have contributed to this book enormously, possibly without them realising it: Helena Howe, Peter Muchlinksi, Mads Andenas, Kateena O'Gorman, Andrea Lista, and Geraint Thomas. Finally, to friends and family who have become used to my extended absences while I have tinkered away in my study. They help me to remember, as is the message in this book, that nothing is either good or bad, but rather thinking makes it so.

Alastair Hudson
Haywards Heath
14 February 2011

Table of cases

Table of statutes

Table of EU Legislation

Introduction

A company is a means to an end. It is an abstract legal device created by human minds. Companies are not found growing in fields, or hanging from trees, or swimming in rivers. They are not a part of nature. There is nothing 'real' about companies in that sense. Nothing can be understood about company law unless we recognise this.

Companies are the products of human minds and human ambitions

Everything that is said and written about companies and about company law is the product of human minds and human ingenuity. We choose to believe that companies exist because the law recognises them as existing and because they are convenient tools. They have become essential to the operation of modern capitalism. Companies are entirely artificial but they are nevertheless very useful in a number of contexts. The whole of company law has been brought into existence as part of our jurisprudence in the United Kingdom both to explain the operation of companies and also to regulate their activities. In this book I will suggest to you that the company device is an empty shell into which human beings pour their ambitions, their aspirations and their activities. There is nothing intrinsically good or bad about any particular company: instead it is a vessel into which capital and work is poured and a conduit through which human activities are carried on. Whether a company is used to achieve immoral ends or productive ends, to raise the common wealth in society or to plunder it, to provide a communal workplace for human beings or to shield assets from taxation, is entirely dependent on the motivations of the human beings who have called that company into existence. As Shakespeare said: nothing is either good or bad, but rather thinking makes it so.

While companies are artificial creations, they nevertheless seem to live

Companies are so much a part of our modern life that it is somewhat counterintuitive to think of them as being merely empty vessels. This is partly due to the fact that most of the companies that are household names are the owners of the large brands which seem to people our high streets, our houses and our workplaces with their gaudy logos and exciting products every bit as much as the human beings who walk there. All companies whose names or logos we recognise will create an image in our minds. For example, when we think of the Nike Corporation, we probably think instinctively of their 'swoosh' logo and their sportswear products. Immediately, we will also have an association in our mind with that name and brand. Perhaps we will think of clean, minimalist stores full of training shoes, or healthy athletes competing on television, or concerns about the possible use of sweatshop labour in the manufacture of those products. What is important is that thinking of that company, the Nike Corporation, will automatically cause us to make an association with its brand. The success of the company's advertising is precisely that we have come to associate that company with something immediate in our brain.

In consequence, companies can seem like living people because they create an association in our minds in the same way that the mention of a human being will also create an association in our minds: based on fondness or dislike or nostalgia or whatever. Companies have therefore come to occupy a particularly important emotional role in our societies. They are part of the warp and weft of life in developed and in developing countries. We should not think that sophisticated branding or corporate presence is solely the preserve of fashionable shopping districts. The global brands reach out to most parts of the world.

In some circumstances, however, people experience companies as their employers (with their letterheaded notepaper and logos), even if in other places they are experienced primarily as manufacturers or as shops or as service providers. The emotional interaction with companies that act as employers is a different one from an interaction based solely on being a consumer of their products. The workplace is a place that provides people with a living wage or community or a sense of self. Importantly, again, the company that acts as an employer stands for something incredibly significant in the human mind as though it actually existed as a distinct person.

Thinking calls companies into life, as though from a blank sheet of paper

Nevertheless, it is important to remember that the company is only a device into which we can pour effort, ideas and investment so that they help us to achieve our goals, to sell our products or to organise our affairs. A company is something that is created.

Imagine a blank piece of paper. With that piece of paper you could do a great many things. You could write a poem on it, or a love letter, or an essay; or you could plan the rest of your life and so keep that letter always around you; or you could write a grocery shopping list on it and discard it in a few hours; or you could write a revision timetable on it and gaze at it regularly in a mixture of fear and awe through the springtime leading up to your exams. The blank piece of paper is a tool for you to use for an infinite number of different purposes. A company is very similar. It is a device that can be customised to suit an infinite number of purposes. A company can be created to be a vehicle for a multinational trading business, or it can be created to hold assets for an individual, or it can be used to operate a charity, or for any other lawful object. This company is in effect a blank piece of paper that can be used for an infinite number of purposes.

In English law a company is a legal person distinct from all other people in the world (*Salomon v A Salomon & Co Ltd* (1897)), as discussed in Chapter 2. That means a company can own its own property, it can make contracts and it can do almost everything in law that an adult human being can do. There are laws governing the way in which a company is brought into being and how it must be operated, but most of those laws provide that a constitutional document governing the company's activities (principally, the 'articles of association') is pre-eminent in deciding how the company is to be run formally. Much of company law restricts particular types of undesirable activity (many of which have their roots in frauds that occurred in previous ages). A large part of company law is concerned with companies providing all sorts of information to the public about their financial circumstances, their directors, their shareholders and so on, to minimise the possibility of fraud. Importantly, company law describes the limits on the liability of the human beings who invest in a company (the shareholders) in the event that the company goes into insolvency, so as to encourage people to invest in companies without fear that they are putting all of their personal wealth in jeopardy precisely because their personal liability as investors will be limited if the company should fail. These ideas are considered in detail in Chapters 1 through 6 of this book: for the moment, we only want to understand the big picture.

At the heart of much company law thinking is a person known as an 'entrepreneur': that is, an individual possessed of insight, energy or commercial acumen who wants to start up a business. A company protects entrepreneurs by limiting that entrepreneur's personal liability to a nominal amount if the business should fail. By offering an entrepreneur limited liability in the event that her business fails, it is hoped that we can encourage entrepreneurs to set up businesses, to take risks and to help our economy to remain vibrant.

That a company is treated by company law as being a legal person in its own right means that a company can create contracts and so on in a way which makes business dealings much more straightforward. Before there were companies with legal personality, it was necessary when creating a contract or transferring land to a business for all of the managers of the business (under partnership law, or agency law, or under trusts law) to be a party to the contract and for dealings to be very complex indeed. It was difficult from the outside to know how the business was organised internally between all of the various business partners. By creating the company, English law made it possible for outsiders to deal only with the company as a single entity, instead of those outsiders (that is, all of us) having to worry about the goings-on inside the company. With a modern company, a contract can be signed on behalf of the company and can be enforceable against the company; property can be sold to a company and bought from a company directly. Furthermore, companies can also use novel ways of raising capital from the public by means of shares and bonds and so forth. All in all, the company offers a much more efficient and attractive way of doing business for commercial people. Company law grew up so as to formalise the ways in which companies were required to operate, as considered in the early chapters of this book.

Companies and the motivations for creating them

Nevertheless, in spite of all the detailed rules that we consider in this book, a company is still a blank sheet of paper on which human beings are able to design their perfect future. An entrepreneur wanting to create a business may choose to create a company because company law will limit her personal liability for any losses from her business to the amount of capital that she chooses to invest in the business herself. If she is lucky or skilful, she may manage to procure investment capital from banks or outside investors so that she has to risk very little of her own money. If the business fails and the company goes into insolvency, the entrepreneur can walk away with her personal property intact. Many captains of industry who are in business today have seen early businesses go into insolvency and they have merely come back to the market again and again with new business ideas until they have had success. This suggests that the company preserves the possibility of entrepreneurs making a success of themselves, whereas if their personal property had been at risk, they might have been ruined by their first business failure.

Alternatively it may make us think that there is potentially something immoral about our company law model if the entrepreneur may bear little or no risk (except for wounded pride) if her business goes into insolvency. We may think that there is 'moral hazard' here: that is, the possibility that people will take unconscionable risks in the management of their companies because they know that they bear little personal risk of loss. We might also be concerned about all of the third parties who traded with that company and who will never be paid what they are owed because the company went into insolvency, even though the entrepreneur is allowed to start up another business after the short period of forced inactivity which is required by insolvency law.

This fear about the ethical problems with limited liability for entrepreneurs may become heightened when we realise that companies are also used simply to hold assets and often to *hide* assets. Most companies in existence do not trade. Instead they exist solely to help organise people's affairs. Suppose Myra owns property which she does not want other people, like Her Majesty's Revenue & Customs, to know that she owns because it generates income on which higher rate tax would be payable. Because a company can own property, Myra could simply create a company and transfer this property to the company. If Myra retains control of the company by means of owning a majority of the shares in it, the outside world may never realise her relationship to that property, but she will nevertheless be able to retain effective control over it by means of her shareholding. This is a very simple example of how people use companies to organise their affairs. The company which Myra sets up could be created in the UK so as to attract a lower rate of tax, or it could be even be established in another jurisdiction where no tax is payable.

It is possible, for example, for a company registered in low-tax jurisdictions like the Cayman Islands to own land and other property in the UK. The tax rules in these situations are too complex for us to consider in this book, but the possibility of manipulating those rules may be the reason for the creation of a company. The more we study company law and notice these things, the stranger that world will begin to seem. Importantly, however, we shall also notice that the potential uses of companies are almost endless.

However, to keep our story simple, let us suppose that the company is established in the UK. There is a legal requirement that shareholdings in companies established in the UK are registered with Companies House, and that the register can be consulted by anybody. Companies are also required to lodge accounts with Companies House. (These detailed rules are considered in Chapter 6.) Therefore, we might think that there is little practical concern about Myra's attempt to avoid tax, nor any risk of her managing to hoodwink her creditors because that register can always be searched. Myra's business creditors could also check up on the company. However, this registration requirement means that the outside world must act as detectives in all of their dealings with companies in that they are required to search the register at Companies House. It may be difficult to know who the controllers of the company are if, for example, Myra decides to use her friend Amanda as a clandestine holder of Myra's shares and as a director of that company, under an arrangement whereby Amanda does whatever Myra asks on payment of a small annual fee. The result would be that it would be very difficult to know that Myra was really in control of the company. In consequence, much of our company law is concerned with the sorts of information which must be made public so as to avoid fraud. Nevertheless, companies can clearly be used simply to achieve clandestine motives in the real world.

Before we get carried away with the idea that companies are only used for reprehensible purposes, we should remember that it need not always be like this. Companies can be formed for very positive social reasons. Most obviously in this regard, companies can be used as charities in England and Wales if they are operated for charitable purposes in the public benefit. Many companies operate businesses that make profits, which therefore pay taxes that fund important social services, and that gainfully employ a large number of people. There are many who would say that operating companies for profit is in itself a good thing because that provides employment for the populace, and the profits they make mean more taxes are being paid, with the result that a virtuous circle of productive economic growth is created. Companies may pursue avowedly positive goals, like The Body Shop did when it began business, as well as naked profiteering (whatever one may think of profiteering, naked or otherwise).

What company law does is to set out the template that companies must use to receive the support of the law. If the members of a company want the support of company law and all of its benefits, then they must follow the rules on the use and operation of companies. As a result, our company law will sometimes be concerned to prevent the abusive use of companies and it will sometimes be concerned simply to set out the formalities that must be performed for a company to be run lawfully. What company law is doing, therefore, is one of two things. Company law may be regulating the use of companies so as to prevent abuse – whether in the form of fraud or abuse of financial markets or whatever – or so as to provide means for compensation or redress for some wrong. Generally, however, what company law is doing is to provide us with a model. Like contract law or express trusts law, company law is saying, 'if you organise your company in the following way, then the courts will enforce your rights, your contracts, and so on'. In that sense, company law offers us a template. Provided we obey the formal rules requiring that we register and operate our company appropriately, then the law has given us a template to use so that we can achieve our goals.

Creation, Frankenstein and the modern company

The analogy which I am going to use to describe this creation of a company in this book is that of Dr Frankenstein. Mary Shelley's novel *Frankenstein* tells the story of Victor Frankenstein, a Genevese student, who discovered the ability to harness electricity so as to give life to inanimate body parts that he had sewn together into a travesty of a human body. What Frankenstein was trying to do was to create something artificially that would appear to act like a person. When we create a company, we are doing something similar. What a person is doing when creating a company is to create something artificially that would appear to act like a person. In English law a company is treated by company law as being a person in the same way that an adult human being is treated as being a legal person.

In the following passage from *Frankenstein*, the young chemistry student had just made a breakthrough when he realised that he could create life artificially so as to mimic ordinary life:

> When I found so astonishing a power placed within my hands, I hesitated a long time concerning the manner in which I should employ it. Although I possessed the capacity of bestowing animation, yet to prepare a frame for the reception of it, with all its intricacies of fibres, muscles, and veins, still remained a work of inconceivable difficulty and labour. I doubted at first whether I should attempt the creation of a being like myself, or one of simpler organisation; but my imagination was too much exalted by my first success to permit me to doubt of my ability to give life to an animal as complete and wonderful as a man. . . . I was encouraged to hope my present attempts would at least lay the foundations of future success.

I would suggest that creating a company gives us this exact same power. We can create anything our minds can imagine. Sometimes we may even create monsters. The company gives us great power thanks to our company law. The company can open a bank account, acquire land, employ human beings, create contracts with suppliers and buyers, and so on. While the company is a blank sheet of paper at the outset, once the company begins to conduct business, then the company *as a company* will appear to have a life of its own which is entirely separate from the human beings who have created it. The original human beings who began the company may die or they may sell the company to other people, but the company continues in existence ostensibly as though nothing had changed.

As I suggested earlier, we have emotional and automatic connections with companies very often in our daily lives whether as customers, employees or simply as people occupying the same lifeworld as those companies. In that sense the company exists separately because we as human beings interact with that company. Our reactions cause its existence. If we ignored Nike and their products, Nike would cease to exist. But the very fact that we have a reaction to the Nike brand and its products means that the Nike corporation exists: whether that is because we covet their trainers as birthday presents or even dislike people who wear their tracksuits as though they were fashionable leisurewear. Our company law causes these companies to exist also precisely because it grants companies legal rights (even human rights in our modern jurisprudence!) and so imposes obligations on the rest of us. Therefore, in that sense, *companies exist*.

Company law causes people to believe in something that cannot be seen or touched. People come to believe that companies own property and can create contracts and so forth because the law tells us that they do. In our modern economies it is the largest public companies that are typically considered to be the most significant actors. Like young Frankenstein, the creator of a company has great power in his hands. He dreams of 'future success' through his creation just as an entrepreneur dreams that a company will create future success for her. What is also important about the company is that it can be as simple or as complicated as the creator wants, just as Frankenstein imagines how complex he could make his creation and how much he could hope to achieve with it. Similarly, companies have been brought into being through the ages by enterprising commercial people to achieve particular goals.

The development of English company law

Company law is a tremendously enjoyable subject to study. Even though it is entirely the creation of human minds, it nevertheless feels very rooted in 'the real world' in the sense that it is the lifeblood of much commerce. We tend to think about commercial life with its phantom companies as being somehow very 'real'. People who argue that we should get rid of this sort of capitalism are often derided for 'not living in the real world'. Yet, there is no proper way of understanding how company *law* is the way that it is unless one understands its history, and you will only begin to understand company law if you recognise how *unreal* companies actually are.

Our modern company law is still based on historical forms of company. The legal ideas behind those early companies have transformed into the enormous corporations we recognise today. You could think of early company law as being a small cottage to which several large extensions have been added over the centuries so that the original cottage constitutes only a fraction of the total size of the building. There has never really been a revolution in the way companies are treated by the law. Instead there has simply been a series of very large extensions built onto the original structure. Company law has developed to meet commercial circumstances throughout the 20th century by means of a number of statutes culminating in the Companies Act 2006 (and many other pieces of legislation besides), which is the longest single statute in the UK. And yet, while that may sound daunting, company law operates on the basis of a few central principles and long-standing ideas.

In essence, as is discussed in the next chapter, companies began life as contracts between people who agreed to form a partnership and to employ other people to act as their trustees in the management of their business. From that seed grew the idea that investors in a company should only bear limited liability if the company were to fail (as opposed to open-ended personal liability to meet the company's debts) and that the company should be seen as a separate legal person (in *Salomon v A Salomon & Co Ltd* (1897)). During the 20th century, a large bureaucracy that had grown up around the administration of companies, the way in which they raise capital, and so forth, has led to more and more law being created to meet modern circumstances. However, that small cottage of traditional legal ideas remains at the heart of the enlarged building. Consequently, understanding the basic layout of the original concepts is essential to understanding the company law we have today.

Ten things you need to know about company law

The reason for including a summary of the ten key concepts of company law

This very brief introductory chapter is intended to give you a list of ten basic propositions about company law which it would be helpful for you to know about before we begin in earnest. Briefly put, if you are aware of these ten concepts, you will probably find the rest of this book much easier to use.

The ten things you need to know about company law

(1) Ordinary companies have limited liability. That means that their shareholders will not be personally liable for the debts or losses of the company. Instead, the liability of the shareholders will be limited to an amount set out in the company's constitution (usually £1) that was the nominal value of each of their shares. This concept was first introduced in the 19th century.

(2) Companies have separate legal personality. This is a result of the decision of the House of Lords in *Salomon v A Salomon & Co Ltd* in 1897. What this means is that a company is recognised by the law as being a distinct legal person, distinct that is from all of the human beings who operate it (such as the company's directors, its shareholders, its employees and so forth). This in turn means that companies can own property, they can create contracts in their own name, they can sue or be sued in their own name, and lots of other things besides. In law, a company can do almost anything a human being can do because companies are considered to be 'people' too. There is an important side effect though. Because companies are legal persons, no human being is responsible for anything for which a company is solely responsible, like a debt. Therefore, even if a company is created by a single human being as a vehicle for conducting her own business, then that human being will not be liable for any legal obligations of the company because the company is a legal person separate from her. Therefore, only the company can be sued (except in limited circumstances which we shall discuss later). It is very rare

for our company law to look behind this so-called 'veil of incorporation' to recognise that the company is simply a device for conducting a business so that a human being can be made liable directly. This also means that companies may be organised into 'groups of companies' which are operated by the same human beings but which are treated by the law as being distinct legal people.

(3)   The principal statute governing company law in the UK is the Companies Act 2006 (CA 2006). It applies to all of the UK as one legal jurisdiction. It is the longest statute ever in the history of this jurisdiction. Nevertheless, it is surprisingly readable because it was drafted for commercial people to be able to read for themselves.

(4)   The directors of companies owe their various statutory, case law and specific constitutional duties to the company itself, and not to its shareholders personally (CA 2006, s 170). The directors of companies are fiduciaries. Their general obligations are set out by the CA 2006, although the statute explicitly recognises that the statutory obligations must be interpreted in accordance with the old case law principles in a delicious combination of the old and the new.

(5)   The powers of the company and the rules that govern how a company operates are set out primarily in the company's constitution (the 'articles of association' specifically after the CA 2006). Because companies began life originally as associations of people governed by the law of contract, modern company law still provides that the articles of association constitute a contract which is binding on the shareholders (CA 2006, s 33). The CA 2006 adds other principles that govern the operation of companies (as discussed in Chapter 6).

(6)   A company's ultimate democratic body is the 'annual general meeting' (or any specially convened general meeting) of shareholders, which may agree that the company can do certain things or that the company can amend its constitution by certain majority votes ('resolutions') identified by the CA 2006. The directors are responsible for the management of the company; whereas the ultimate ownership of the company is said to reside with its shareholders (even though the shareholders do not own the company's property directly). The shareholders do have statutory powers to replace the directors (CA 2006, s 168) at the annual general meeting. In practice then, particularly in relation to large public companies, the interrelationship between shareholders with large shareholdings and the directors will be a particularly sensitive one.

(7)   Shareholders may not sue on behalf of the company. If a company suffers a loss, then only the company can act as a claimant to sue to recover that loss, unless the shareholders have received permission to bring a 'derivative action' on behalf of the company (CA 2006, s 260). If the company has acted in accordance with its constitution, or if a majority of the shareholders have voted for a course of action, then minority shareholders may not object to that

action unless they can demonstrate that it is 'unfairly prejudicial' to them (CA 2006, s 990). Otherwise, shareholders can seek to have the company wound up where the court is convinced that it is just and equitable to do so (Insolvency Act 1986, s 122). These shareholder rights are analysed in Chapter 8.

(8) There may be different classes of shares in a company: ordinary shares, preference shares, convertible securities, and so on. Ordinary shares grant the shareholders one vote per share at a company meeting; they also grant their shareholders a right to receive a dividend (in the same amount for each share of the same class) if the directors decide that there are sufficient distributable profits that can be divided among the shareholders. Importantly, though, the shareholders do not have property rights in the company's property[;] except that if the company is wound up, they have a right to receive a proportionate share of the company's property. By contrast, the owners of preference shares usually have a right to receive a fixed dividend from the company. The share-holding of a company may be divided between different classes of shareholders. The rights of shareholders within a given class of shares will be identical to one another, except that owning more shares acquires proportionately larger rights to receive dividends or to exercise more votes at a general meeting.

(9) There are two types of companies which are particularly important for our purposes: private limited companies (which must have 'limited' or 'ltd' after their name) and public limited companies (which must have 'public limited company' or 'plc' after their name). Private companies may not offer their shares to the public (CA 2006, s 755). It is only the public companies which may have their shares offered to the public, for example by being traded on the London Stock Exchange. The issue and trading of the shares of public companies on regulated markets like the London Stock Exchange are governed by the securities regulations considered in Chapter 11. There are general principles governing the transfer of and dealing with the shares of all companies, as discussed in Chapter 10.

(10) There are numerous restrictions on dealings with, or alterations to, a company's capital so that the position of third parties is not adversely affected. This includes limitations on giving financial assistance to people to buy shares in the company, or buying back or cancelling shares, or paying dividends out of the company's capital, or restructuring the company's capital. There are detailed statutory and case law principles governing the way in which such things can be done in limited circumstances, as discussed in Chapter 10.

How this book can help you to understand company law

So, there are the ten things that I think it would be really useful for you to know at the outset about company law so that this book will make more sense if you are determined to read it in a hurry, tearing through its pages seeking knowledge like the hordes of an invading army pillaging the countryside. If you are not in such a hurry, then knowing those ten things will nevertheless make reading the rest of this book much easier.

Ideally, before your course begins, you will be able to afford to set aside the three or four hours that it might take to read this book from cover to cover in one go (including coffee and text message breaks), in which case you will see that each idea is explained patiently as it arises. However, I am also aware that other readers will be in a hurry to ingest information at high speed because exam day is looming, or because coursework must be handed in, or because the pressure is too much and you need something to help you halfway through. So, this book has been written both as a single book to be read from cover to cover, expounding a central thesis, and also as a book which you can dip into from time to time. Nevertheless, if you read a textbook even in outline before lectures, your work in lectures is more informed than it would be otherwise, and consequently your notes and your preparation for seminars and exams will improve markedly.

This book is in fact written as an essay about the nature of company law. It begins with the opening metaphors of Frankenstein's monster and of the company as a fictitious, abstract device that helps us to achieve very real objectives. The use of these metaphors throughout the text, together with observations about some key moments in British history that impacted on company law (such as the South Sea Bubble and the Industrial Revolution powering the Victorian British empire), will help to explain a number of seemingly abstract principles of UK company law which you will need to know about but which may otherwise seem impenetrable. After all, if something does not make sense, the easiest means of understanding it is to ask why it was put there in the first place.

By understanding the genesis of the company and the purpose of some of the key principles, we will be able to reflect on the usefulness and desirability of those rules in the modern world. Successful essay-writing for students involves the use of metaphor and 'big picture' ideas to bring formulaic discussions of the law to life. This little book is packed with those sorts of ideas, and suggestions of avenues down which your work in company law might progress. For example, the later chapters on 'Corporate social responsibility' and 'The future of company law' are intended to deal with exactly those sorts of ideas, many of which are staples of the more progressive company law courses.

Well, I must just say how much I have enjoyed this time we have spent together. Nevertheless, I think it is time we got down to work. We will begin with an outline of the history of company law, before then considering the foundational decision of the House of Lords in *Salomon v A Salomon & Co Ltd* in Chapter 2.

Chapter 1

The birth of modern company law

Introduction

This book analyses the foundational principles of company law in detail. It is only possible to understand many of those principles if we understand their historical and jurisprudential bases. For example, it is only possible to understand the nature of the duties of directors if one understands how the law conceived of directors in their earliest incarnations as trustees. Much of the modern law will make more sense if we understand its roots. Similarly, it is only possible to unpick many of the problems with modern company law if the history of the company is understood. So, in this chapter we begin, as all good stories do, at the beginning.

The meaning of a 'company'

The word 'company' is derived from the Latin words 'com' and 'panio' and gave rise to the word 'companio' (from which we also derive the English word 'companion') meaning literally 'someone with whom you break bread'. Therefore, in its earliest usage, a 'company' was a group of people who ate together. If you mix in polite society you might occasionally belong to a 'company' of people going to the theatre, or you may form part of a 'company' having dinner together. A good friend is someone we might think of as being 'good company'. These are the older definitions of the word 'company'. Therefore, the concept of a company began as an agreeable collection of people joining together for some social or joint purpose. A company was simply an association of people in its earliest form.

The birth of the joint stock company

In time companies also came to represent people who banded together specifically to trade. Literally, the traders formed a 'company' of people with interests in common with one another. It is from this root that our legal conception of the company grew. The earliest companies of that sort can be identified most clearly in England in the mediaeval town corporations and, more familiar to our commercial understanding of the company today, in the Elizabethan era when the trading nations of Europe were seeking new lands which they could colonise, where they thought they could collect treasure and with whose people they could trade. In truth, Europeans had been doing this for centuries, just as Christopher Columbus had done in 1492 when he reached America.

It was in the sixteenth and seventeenth centuries that the idea of the *joint stock company* can be understood as having begun. In the simplest example, a number of rich English merchants and traders would acquire the use of a ship, which they filled with their different items of stock. In that sense the boat contained their 'joint stock'. Someone was appointed to act as captain of the ship. The captain's job was to sail the ship to whichever destination had been chosen and to trade the joint stock for the fabulous wealth that was believed to exist in faraway lands. The 'South Seas' were a popular destination at the time – what we now call South America – because it was believed that gold lay on the ground there, once one had sailed past sea serpents and dragons, and that this plentiful gold could be acquired in exchange for the stock on the ship. (In effect, somewhat optimistically perhaps, it was hoped that a few sheepskins could be exchanged for huge quantities of gold.)

In legal terms, the traders formed a contract between themselves, which we would analyse today as being a partnership. That means that the traders agreed to share the profits and losses of this venture between them in accordance with the terms of their partnership contract. The ship's captain was understood as being a person who had legal powers to deal with the stock held on board the ship just like a trustee, and who also owed the fiduciary duties of a trustee to the traders to sell their goods for the best price and to bring the profits home. The captain owed fiduciary duties to act fairly and even-handedly between all of the traders just as any trustee would. The traders were therefore not only partners in contract law but they were also beneficiaries of this trust arrangement under trusts law. The trust property was the stock held on board ship, and any property or money for which it was exchanged on arrival in the South Seas. Therefore, the joint stock company was an imaginative use of the centuries-old ideas of contract, partnership and trust to achieve a very particular relationship that permitted trade to be conducted overseas. This was how the commercially useful idea of the joint stock company developed.

From illegality to necessity

The earliest companies were, however, beset by fraud. Knowing a little about this early history will help to explain a large amount of company law, from the general duties of directors to the detailed rules governing the maintenance of a company's share capital. In the early 18th century, a man called John Law had the exceptionally clever idea of encouraging people to exchange their gold and silver coins for paper money so that the French king could control the amount of money in circulation in France. This also created the possibility that money could be raised from the public more easily to wage wars or expand French trade. A company was formed in Paris to trade in America, called the Mississippi Company. In 1719 in Paris, speculative fever was at its height as people fought to give their money over to the Mississippi Company to acquire a 'share' in the riches, which everyone agreed was sure to be earned by members (or shareholders) of the company from the company's prospective trading activities in America.

In 1720 in England a man called John Blunt had a similar idea. A century of waging war had left the English monarchy in debt. Money had to be borrowed from the public and those investors had to be paid interest on their loans as well as being paid their money back. When the new king, George I, had acceded to the throne, he was a stranger in the country he ruled because he was a German import needed to succeed to the thin Protestant bloodline of English royalty. Saddled with a huge national debt and few friends, the new king and his government fell into Blunt's thrall. At that time the state borrowed money by issuing bonds, in return for which the bondholders were entitled to receive interest. Blunt's idea was to encourage people to give up their bond rights in exchange for shares in the South Sea Company. The South Sea Company purported to trade in the South Seas and promised to generate unimaginable wealth for those fortunate enough to hold shares in it. A speculative fever spread in London just as it had spread in Paris. Everyone, from members of the royal family down to the petit bourgeoisie wanted to own shares in these new companies. It was a common belief that their investments could not fail. Indeed, in London each new issue of shares was greeted with tremendous public excitement.

Trading in these shares in the London coffee houses around Exchange Alley was feverishly excited. Huge fortunes were made in the short term in buying and selling shares in the South Sea Company. But, in truth, John Blunt was operating a sham. He had loaned money to well-placed individuals from the company's funds to buy shares in the company so that it appeared that the company's shares were much sought after. This is known today as 'making a market' in shares and is a criminal offence ('market manipulation', see Chapter 11). In fact the company had no business and all of this share speculation went on before a single ship had sailed. People willingly exchanged their bonds against the government, which entitled them to be paid money by the State, for shares in an illusory company. When the fraud became public knowledge, many people were ruined, including members of the royal family and Cabinet ministers. As a result, companies were made illegal for about a century because they were associated with fraud and immoral speculation. (The South Sea Bubble is discussed brilliantly by Balen, 2002.)

By the 19th century and the Industrial Revolution, however, there was a need for more investment capital to fund and fuel economic expansion. The construction of the railways and the infrastructure of modern English towns (for our sewers and so on still rest on Victorian ingenuity, muscle and skill) needed to be funded. To do this, companies were legalised again in 1824 so that investment capital could be raised from the public, instead of relying on ordinary bank borrowing, which could not meet the pace of growth.

It is important to note, however, that companies have a heritage and a root in English law as being based both on commercial expediency and the possibility of systematic fraud. Much of modern company law is concerned to encourage the former while avoiding the latter. This explains a large number of the rules underpinning company law today because they were created in the light of fraudulent schemes. Cases like *Gluckstein v Barnes* (1900) illustrated the sort of fraud with which early companies were associated. It was a commonplace of Victorian fiction that companies and commercial people were often engaged in fraud: examples are the Anglo-Bengali Disinterested Life Assurance Company in Charles Dickens's *Martin Chuzzlewit*, Melmotte's specious railway company in Thackeray's *The Way We Live Now* and financial skulduggery in the City of London in Frederick Wicks's *The Veiled Hand*.

In *Gluckstein v Barnes*, four men promoted a company that bought the Olympia exhibition premises in London. They raised investments from the public on the basis of the plausible-looking prospectus that they issued. The prospectus suggested that all of the investors had paid in full for their shares, when in fact the promoters had acquired their shares at a discount. When third parties had paid for their shares, the company was allowed to fold once the promoters had redeemed their shares for their full face value. Lord Macnaghten explained both how well established this sort of fraud had become and how exactly the transaction worked:

> It is the old story. It has been done over and over again.
>
> These gentlemen set about forming a company to pay them a handsome sum for taking off their hands a property which they had contracted to buy with that end in view. They bring the company into existence by means of the usual machinery. They appoint themselves sole guardians and protectors of this creature of theirs, half-fledged and just struggling into life, bound hand and foot while yet unborn by contracts tending to their private advantage, and so fashioned by its makers that it could only act by their hands and only see through their eyes. They issue a prospectus representing that they had agreed to purchase the property for a sum largely in excess of the amount which they had, in fact, to pay. On the faith of this prospectus they collect subscriptions from a confiding and credulous public. And then comes the last act. Secretly, and therefore dishonestly, they put into their own pockets the difference between the real and the pretended price. After a brief career the company is ordered to be wound up. In the course of the liquidation the trick is discovered.

Interestingly, this sort of skulduggery was considered to be an old story even in 1900. The investing public is conned by means of being fed false information. Nevertheless, the pace of economic growth in the Victorian era meant that companies were encouraged to proliferate as a means of raising money from the public, and so company law had simply to develop rules that tried to prevent fraud while allowing companies to grow. One particularly significant development was the advent of limited liability, which is considered next.

Limited liability

One of the key developments in the growth of the company was the development of *limited liability*. An investor in an early company faced the risk that she would be personally liable for all of the company's debts in the event that the company went into insolvency or that it failed to make enough profit to pay off its creditors. Under partnership law, for example, unless the arrangements had been very cleverly organised, each partner would be personally liable for all of the debts of the business. So, partners and therefore all investors in companies ran the risk of losing all of their property if the business failed. This was a huge disincentive to investors. So, in 1844 a statute was passed which permitted companies to award limited liability to their shareholders. This meant that a company could be created by investors paying an amount of money agreed between them and set out in the company's *constitution* as an initial investment, which would form the company's *capital*.

That the investor has limited liability means two things. First, the investor as shareholder is not personally liable for the company's debts (unless she agreed separately to guarantee the company's debts). Second, the investor is only risking the loss of her original investment in the company (i.e. the purchase price of the shares), but is not risking an open-ended exposure to the company's future losses. On the upside, the investor could receive a share in the company's profits proportionate to her shareholding. This encouraged more investors to invest in companies, particularly given the pace of economic growth in Victorian England as not only did the companies build railways, bridges and viaducts that were the engineering wonders of their time, but the British also established one of the largest empires ever known across every continent, which created further trade and ever greater wealth (whatever one may think of that). Consequently, the combination of limited liability and the conspicuous generation of great wealth through the marvels of the Victorian age encouraged more and more people to invest in companies.

The joint stock company became a different animal. Professional managers operated the companies not simply as servants of rich aristocratic patrons and traders, but rather as a growing bourgeois class in their own right. The new age changed the British class system irrevocably, as people who had formerly been serfs in the fields moved into the dangerous, dirty new factories in the new towns effectively as slave labour and, if they were lucky or clever, as part of a new petit bourgeoisie which could earn a living with its brains instead of its brawn. The empire offered prospects for imaginative men (mainly) to make their fortune. The company was an important part of spreading the fruits of this activity by spreading the possibility of investment gain among the bourgeoisie. It also created a larger class of bourgeois professionals who worked in the banks, around the Stock Exchange, and in law and accountancy offices.

It is sometimes said that most of modern history is actually the history of the growth of the bourgeoisie, and there is much in that: the upper and lower classes stay in broadly similar conditions, whereas the change and movement happens primarily in the middle. Social change in Britain has had a lot to do with the steady dismantling of monarchical and aristocratic power such that most power in modern society is held uneasily between companies and 'the markets' on the one hand, and government and the State on the other.

All of this discussion of breathless growth and of limitations on investor liability has catapulted us to the centre of the modern company as a means of effecting trade, but it has overlooked two important dimensions in the development of the concepts of company law: the roots of companies in the law on associations and the development of the idea of corporate bodies in the corporations formed by royal charter.

Associations, and their links with modern companies

Modern companies, as understood by English jurisprudence, began in the law of unincorporated associations. Anyone who has studied property law at an English or Welsh university should be aware of the law on unincorporated associations. Associations are clubs or societies or groups of people that are not organised as companies. In that sense they are said to be 'unincorporated', whereas a company is a body that has been *incorporated* (see Chapter 6). When companies were unlawful, people still wanted to group together. They did so lawfully in the form of associations, which had long been a part of English law. An association is a contract between a group of people who want to carry on an activity in common. A good example of such an association is a student law society at university. A well-organised association will have a *constitution* that sets out its rules in contractual form. Typically, members will pay a subscription to join the association (just as an investor in a company buys *shares* at the outset so as to become a shareholder in that company). This contribution stands for the contractual 'consideration' that is necessary to make an English contract valid. The constitution will deal with questions such as the powers of the various officers of the association, the election and dismissal of those officers, what happens to property that people give to the association (such as subscriptions or legacies), the members' rights to vote so as to control the activities of the association democratically, the rights of the members to receive accounts as to the association's property and activities, the circumstances in which the association will be wound up, and the rights of the various members to take property rights in the association's property after it has been wound up.

The joint stock company and the modern company are built on these foundations. Modern companies have members. They are more colloquially known as 'shareholders' in limited liability companies, although they are commonly described as 'members' in the companies legislation, just like members of an association. Company law sets out the rights of shareholders if something goes awry, and company law specifies which matters need to be decided by a voting procedure, as discussed in Chapter 6. However, while company law presents a default setting for companies with inadequate constitutions and deals with many types of abuse within companies, it is the company's own constitution which specifies the company's objectives, the powers of the directors, the rights of shareholders, and many, many other points of detail concerning the operation of the company, as discussed in later chapters. A company's constitution is known formally as its *articles of association* today. There was once another constitutional document, known as the *memorandum of association*, which set out the company's objects, but the memorandum of association is no longer required after the enactment of the Companies Act 2006 (CA 2006), although some companies still have them. These various constitutional documents are discussed in Chapter 6.

What is important to understand at this stage is that the somewhat peculiar-looking articles of association purport by s 33 of the CA 2006 to create a contract between the shareholders when those shareholders may never have met, let alone created a contract as to all of the complex provisions in the articles of association. However, this contract makes complete sense when we understand it as being simply a modern evolution of the old contract between the members of an unincorporated association as contained in its constitution, as has been the practice of the law of associations for centuries, which has remained in our company law. Company law still uses the expression 'constitution' to this day.

Let us think of a university student law society as a clear example of an unincorporated association. Members join and leave a student law society all the time, and the constitution binds them as a contract without any difficulty. New joiners are bound by the same contract as binds long-standing members. The constitution of a company operates in the same way, constituting the rules by which that company operates without the parties needing to say anything more between them. The rights of the original members of a company when it is created and the rights of people who become shareholders later will be governed by the articles of association (as well as by company law). In an unincorporated association, the constitution and the officers in charge of the association decide how to run the association, how to allocate its property and how best to pursue the association's goals. The association's constitution explains how the members are able to object to the officers' decisions, and what rights they have to share in any profits. Similarly, in a company it is the management (primarily the directors in most companies) who manage the company and the members (as shareholders) have specified rights to vote at annual general meetings, to receive accounts, to vote out the directors if they wish, and to receive a distribution from any profits by way of a dividend. Clearly, modern companies have emerged from the fertile soil of the law of associations, which is itself a combination of contract law and property law ideas.

Corporations, and the growth of legal personality

While we have focused on the Victorian enthusiasm for companies, the idea of a corporate entity was not one created by commercial activity in the 19th century.

The notion of a corporation has a much older provenance than that. The first companies to be formed in English law were the product of the municipal corporations of the 13th century. The feudal system did not apply evenly across England. Rather, a number of towns remained subject to direct monarchical control subject to the terms of their royal charters. Out of these charters grew an understanding that the people of the town were more than merely individuals but rather constituted a collective entity. These municipal entities formalised their rules within the scope of their applicable royal charter and acquired 'older men' (or the more familiar 'aldermen') and wardens as their officers. In parallel, trade 'gilds' (or guilds) were also formed as corporations – most notably the 12 grand liveried companies in London. These early corporations saw conflicts common to the modern company. For example, in the Founders' Company there was a dispute between the liveried members (who quite literally wore liveried coats and who carried most of the control over the activities of the company) and the yeomen (effectively junior members of the same company) as to prices fixed by officers of the company. The Court of Star Chamber intervened in favour, ultimately, of the yeomen in part (*Butlond v Austen* (1507)). Effectively this was a conflict between senior and junior members of the corporation as to that corporation's pricing policy. This mirrors the disputes between majority and minority shareholders in modern company law today.

Out of the pomp and circumstance of liveried companies and town corporations, and the emerging notion of collective entities through which the people could act, grew the company. The company was an expression of the personality of those individuals acting together. The joint stock company evolved as a commercial undertaking that (subject to the historical controls on companies considered later) was made up both of the 'joint stock' provided by the members and also of the 'company', with the two elements being separate issues of property and of contract until the jurisprudence of the late 19th century. Even in the famous decision of the Court of Appeal in *Smith v Anderson* (1880) there remained an explicit understanding of the company as merely an expression of the common endeavour of its members and not as an entity with its own personality. While it may seem remarkable now, the unincorporated companies of the early 19th century were considered to be illegal contracts by Lord Eldon (*Josephs v Preber* (1825)). It was not until 1843 that the idea that companies constituted illegal, and therefore unenforceable, contracts was displaced in the common law (*Garrard v Hardey* (1843); *Harrison Heathorn* (1843)).

Then in 1897 everything changed. A decision of the House of Lords in *Salomon v A Salomon & Co Ltd* (1897), to the effect that a company should be recognised as being a distinct legal person from the human beings who had created it, changed company law for ever and set in train a fundamental reorganisation of commercial activity around the world. So, even though Aron Salomon had traded as a bootmaker in Whitechapel for many years, when he incorporated his existing business by forming a company and transferring his business to the company, but otherwise continuing to trade much as before, it was held that the business's creditors could only proceed against the company and not against Aron Salomon personally. From this simple notion grew the most important idea in the whole of modern capitalism. Its legacy has been the rapid spread of capitalism across the entire planet using companies with separate legal personality as it is standard-bearer. That is what we consider next.

Chapter 2

The *Salomon* principle

Introduction

In the previous chapter we considered how the modern company grew of out of the law on unincorporated associations, how it used ideas long identified with town corporations created by royal charter, how it evolved from the joint stock company, and how shareholders in companies were granted limited liability by statute. One key element of the modern company, however, remained outstanding: the principle of separate corporate personality, which was created by the House of Lords in *Salomon v A Salomon & Co Ltd* (1897). We will refer to this principle as 'the *Salomon* principle'. We will begin with a close reading of the *Salomon* litigation.

The *Salomon* litigation

The history of Aron Salomon, and the background to the legislation

Aron Salomon was a leather merchant, hide factor and boot manufacturer on Whitechapel High Street in London's East End. One of his principal customers was the British Army. Aron had traded successfully as a sole trader. That he was a 'sole trader' at first means that he owned all of the business and all of its assets, and that he employed the workforce directly himself. (He was also a 'sole trader' in that he made shoes!) However, he was well advised and so rearranged the structure of his business – even though to the outside world it must have seemed that everything carried on exactly the same as before. The restructuring involved Aron creating a company. At the time the legislation (the Companies Act 1862) required that there were seven 'members' (or shareholders) of the company, and in this particular instance a trustee who held the company's property on trust for it. Therefore, Aron organised that the members would be himself, his wife, and five of his adult children so as to comply with the legislation in effect at the time. What this achieved was a shift in the risk of the business failing off Aron Salomon personally and instead onto the company (and thus its creditors).

Once Aron Salomon created a company, he transferred all of his business assets from his personal ownership as a sole trader to the new company called A Salomon & Co Ltd. Notably this company even bore the name of the man who continued to run its business. I have a very clear picture in my mind of a brick-built Victorian factory in Whitechapel with wrought-iron gates opening onto a yard in front of the main factory buildings. I have no reason to know that this is how the factory actually looked, but it is my mental image. However, in my mind's eye I imagine the only visible change to the outside world once that company became the owner of the business was a change to the sign over the wrought-iron gates in front of the factory, so that it no longer read 'Aron Salomon, Bootmaker' but instead probably read 'A Salomon & Co Ltd'. To all intents and purposes, however, the rest of the business carried on the same way: Aron Salomon would have arrived at the same time every morning, everyone would have treated him as the man in charge, and customers would probably have paid the same prices for the same product. (Little is known about the detail of Aron Salomon's business, although an excellent account of what is known is set out in Rubin, 1983.)

The very clever idea which was at play here was that Aron Salomon no longer bore personal responsibility for all of the potential losses of his business. Instead, once the company had been created, the business belonged to that company and it was therefore hoped that if there were any losses in the future, it would be the company that would have to bear those losses while Aron Salomon sat snugly at home in his parlour. And yet the only real change to the way in which Aron Salomon conducted business, so as to comply with the formalities of company law, was that he was required to have meetings of the company in which his fellow shareholders would have to vote on those matters requiring their agreement in the company's constitution. Interestingly, Aron Salomon structured the shareholdings so that he could outvote all of the other shareholders. Consequently, he was still as in control of the business as he had been beforehand. He was still the boss and the business continued as before. In any event, the other members of the company were all members of Aron Salomon's immediate family.

It was a change only of form and not a change of substance. The problem was that, before Aron Salomon's business structure came before the courts, it seemed that the legal community generally (and certainly the judiciary) had not thought of using the company in this way; rather, it was only clever practitioners of the sort who advised Aron Salomon who had had this idea. As ever, it is the ingenuity of ordinary lawyers in practice that develops so much of our common law. The aim of the company legislation had been that seven or more unrelated human beings would come together for business purposes and limit their liability by forming a company in accordance with the Act. It had not been imagined that a sole trader would be able to reorganise his own business so as to attract the protection of limited liability by getting six of his immediate family to act as his stooges.

In time, business became difficult for A Salomon & Co Ltd and eventually the business failed. To borrow money for the business, Aron Salomon had granted debentures (a security involving a sort of charge, discussed in Chapter 10) over the company in return for loans (from Mr Broderip among others). In essence, this meant that the lenders had charges over the company's property. When the company went insolvent, the lenders sought to recover their money from Aron Salomon personally, and in so doing they were able to sue on behalf of the company (A Salomon & Co Ltd) under the niceties of company law: hence the case being Aron Salomon, the human being, versus A Salomon & Co Ltd, the company. By the time the case reached the House of Lords, with a sort of maudlin inevitability, Mr Salomon was described as being 'Aron Salomon, Pauper'. This meant that he had become bankrupt. We must remember the stigma that attached to poverty in Victorian Britain: there were debtors' prisons (like the Marshalsea Prison immortalised in Charles Dickens's *Little Dorrit*) that imprisoned bankrupts until their debts were paid. This was why Aron Salomon was referred to in the Law Reports, as though it was some sort of qualification, as a 'pauper' then as opposed to a 'bootmaker'.

What we shall do now is to consider each of the stages in the litigation in turn. The High Court and the Court of Appeal considered, in effect, that what Aron Salomon had done in setting up a company in this way was an abuse of the companies legislation; whereas the House of Lords took a radically different approach and assumed that the company was itself a distinct person at law, with the result that Aron Salomon was not personally liable for the company's losses. The question was this: could Aron Salomon claim successfully that the obligations under the debentures were owed by the company alone and therefore that Aron Salomon had his personal liability as a shareholder limited as a result of the Companies Act 1862?

The decision at first instance: Broderip v Salomon (1895)

When the litigation came on before the High Court, the judge was Vaughan Williams J, who was a very experienced judge and the author of a celebrated book entitled *Williams on Bankruptcy*. It expresses, if you like, the old morality about the conduct of business. Vaughan Williams J was clearly concerned that allowing an arrangement of this sort to be enforceable so that Aron Salomon's liability was limited would have the effect of robbing the business's creditors of the chance of recovering their money. It should be recalled that as an authority on bankruptcy law, Vaughan Williams J would necessarily be concerned about the proprieties of protecting creditors against the risk of loss when the business went into insolvency. His Lordship recognised the problem that 'the company was a mere alias of the founder. In this case it is clear that the relationship of principal and agent existed between Mr Salomon and the company.' Interestingly, his Lordship did accept impliedly in the second sentence that the company was capable of being a separate person who could act as agent for Aron Salomon (and therefore must have been a distinct legal person – an idea which is seized upon by the House of Lords). Nevertheless, Vaughan Williams J held that Aron Salomon should not be able to rely on the limited liability provisions of the companies legislation.

The decision of the Court of Appeal (1895)

The Court of Appeal contained very experienced commercial judges, not least Lindley LJ who was the author of *Lindley on Partnership*, which is still in print today. Lindley LJ began his judgment by recognising that there had been a growth of 'one-man companies' in the years immediately leading up to this decision. Clearly the Court of Appeal considered not only that what Aron Salomon had done was contrary to the spirit of the companies acts but also – more significantly perhaps – that there was something immoral about it. As Lindley LJ held:

> There can be no doubt that in this case an attempt has been made to use the machinery of the Companies Act, 1862, for a purpose for which it never was intended. The legislature contemplated the encouragement of trade by enabling a comparatively small number of persons – namely, not less than seven – to carry on business with a limited joint stock or capital, and without the risk of liability beyond the loss of such joint stock or capital. But the legislature never contemplated an extension of limited liability to sole traders or to a fewer number than seven. In truth, the legislature clearly intended to prevent anything of the kind . . . Although in the present case there were, and are, seven members, yet it is manifest that six of them are members simply in order to enable the seventh himself to carry on business with limited liability. The object of the whole arrangement is to do the very thing which the legislature intended not to be done; and, ingenious as the scheme is, it cannot have the effect desired so long as the law remains unaltered. This was evidently the view taken by Vaughan Williams J.

That the Court of Appeal considered this structure to be immoral emerges from the following sentence from Lindley LJ. Having observed that the company had seemingly been properly created following the appropriate procedure, his Lordship held that:

> The company must, therefore, be regarded as a corporation, but as a corporation created for an illegitimate purpose.

Later on, his Lordship considered that the company had been 'improperly created' and that the other shareholders had no genuine interest or role in the company. The morality is clear: this sort of clever use of the legislation was 'illegitimate' because it was merely a sham intended to protect Mr Salomon from liability for the costs of his business, and so would not be supported by the courts. Instead, the rights of third parties and the proprieties of dealing with one another in a straightforwardly honest way would be required. (It seems to me personally that this is in effect a judgment that is suggesting that Aron Salomon was taking unconscionable advantage of the legislation as in the old equitable maxim that 'statute cannot be used as an engine of fraud'. I will return to that idea later.)

As Lindley LJ observed:

In a strict legal sense the business may have to be regarded as the business of the company; but if any jury were asked, Whose business was it? they would say Aron Salomon's, and they would be right, if they meant that the beneficial interest in the business was his. I do not go so far as to say that the creditors of the company could sue him. In my opinion, they can only reach him through the company.

Ultimately, we have the following bald statement from the same judge:

Mr Aron Salomon's scheme is a device to defraud creditors.

There is no doubting his Lordship's view of the purpose and the morality of this scheme: it is nothing less than fraudulent in his Lordship's view. Lopes LJ held that: 'It would be lamentable if a scheme like this could not be defeated' because it is merely a veil behind which Aron Salomon was hiding from his creditors. Note that word 'scheme' is used throughout the Court of Appeal's judgments: this is not what a company was supposed to be under the legislation; instead it was a clever scheme, scam or charade to elude creditors. And, gloriously in tune with the thesis of this book, Lopes LJ held that if Aron Salomon were allowed to get away with his scheme, the Court of Appeal would 'be giving vitality to that which is a myth and a fiction'. The modern company is just that: a legal fiction that has commercial usefulness. It is a fiction in the sense that it is not a tangible person, but the law nevertheless pretends that it is a legal person.

What is interesting is that the House of Lords decision (considered next) ignored the moral arguments and instead focused on the illogicality of Lindley LJ, accepting that the creditors can only reach Aron Salomon *through* the company, and instead recognised that this meant that the company was itself a distinct person. There are those in the 21st century who still maintain that the whole notion of a company is simply a means of cheating third parties out of their money and distancing the shareholders from any personal responsibility. Lindley LJ has a scent of this when he held that companies:

are mere devices to enable a man to carry on trade with limited liability, to incur debts in the name of a registered company, and to sweep off the company's assets by means of debentures which he has caused to be issued to himself in order to defeat the claims of those who have been incautious enough to trade with the company without perceiving the trap which he has laid for them.

As Kay LJ held, the companies acts 'were not intended to legalize a pretended association for the purpose of enabling an individual to carry on his own business with limited liability in the name of a joint stock company'. On this view, what Aron Salomon did was outwith the purpose of the legislation.

It is telling that the company is said to be a 'device', a 'scheme' and a 'trap'. The approach taken in the Introduction to this book (which is part of the core argument of this book) is that there is nothing necessarily wrong or immoral in the modern company, but rather that it is possible to use the modern company in ways that may have immoral outcomes. Let us see how the legal understanding of the modern company came about in the House of Lords.

The seismic impact of the decision of the House of Lords: Salomon v A Salomon & Co Ltd (1897)

In essence, the House of Lords held that the logical result of the company acting as the agent of Aron Salomon was that the company was itself a person; and if the company was itself a person, then the law should treat all of the assets and liabilities of the business as belonging to the company as a distinct legal person as opposed to belonging to Aron Salomon. It is no exaggeration to say that at that moment the whole of modern capitalism was born. It would not have been possible for global markets to have developed in the way that they are today if the convenience of contracting and raising capital through companies had not been developed. Of course the House of Lords could not have seen the world to which they were giving birth – for them, the only immediate effect would have been the protection of sole traders and others from liability for a business's losses. However, the notion that a company exists as a legal person meant that a new commercial beast was born. In the ether around the world, businesses became people. They rose up and walked. Like Dr Frankenstein conducting electricity through his monster, the ideas of the House of Lords produced a form of life.

What is remarkable is that the House of Lords cited no authority, nor any economic or political thinker, to support this development. There is no reliance on Adam Smith's *Wealth of Nations*, or anything of that sort. Instead, the birth of capitalism results from an exercise in logic and, purportedly, statutory interpretation. It is a good example of the law of unintended consequences. As Lord Halsbury put it, perhaps without much argument:

> it seems to me impossible to dispute that once the company is legally incorporated it must be treated like any other independent person with its rights and liabilities appropriate to itself, and that the motives of those who took part in the promotion of the company are absolutely irrelevant in discussing what those rights and liabilities are.

Notice the words 'it seems to me'. There is no pre-existing authority. Instead, Lord Halsbury was relying on his own view of logic. As a result, the birth of capitalism happened by accident. The logical argument Lord Halsbury used to refute Vaughan Williams J was as follows:

> I confess it seems to me that that very learned judge [Vaughan Williams J] becomes involved by this argument in a very singular contradiction. Either the limited company was a legal entity or it was not. If it was, the business

belonged to it and not to Mr Salomon. If it was not, there was no person and no thing to be an agent at all; and it is impossible to say at the same time that there is a company and there is not.

In refuting Lindley LJ it was held that it was not possible to say that this was contrary to the intention of the legislation when it was not possible to know from the words of the legislation what the purpose of the legislation was. (This was at a time when the courts were not permitted to consult parliamentary debates to identify the purpose of legislation.) As a result, the company was found to be a distinct legal person, and therefore Mr Broderip and the other creditors could only recover their losses from the company and not from Aron Salomon personally.

We have spent a large amount of time on this one particular case because it tells us something quite remarkable. It tells us that the central tool of modern capitalism, the company with legal personality, came into existence without any discussion of its source or meaning. Instead, the logic that appealed to one elderly man in particular (Lord Halsbury) sowed the seed of the economic organisation of the developed world for the next century and more. From the human mind we developed an idea of legal personality, but the detail of capital markets raising money for companies and of shareholders in clothing companies being able to keep themselves at many removes from the sweatshops producing their goods would be developed by clever practitioners over time. Just as Aron Salomon's advisers had been clever enough to realise that the Companies Act 1862 could be used in ways the legislators had never imagined, the development of the company and the reaction of company law to it would also be the product of the ingenuity of corporate lawyers.

The effect of the Salomon litigation on company law

One thing that should strike any reader of the *Salomon* decision, however, is the evident certainty among the members of the House of Lords that they were right. Their Lordships' speeches make little reference to any pre-existing case law, nor to any contemporary non-legal ideas or sources. Given the determination of the Court of Appeal in *Smith v Anderson* only 18 years previously that the company remained a species of trust, and the views of Vaughan Williams J and the Court of Appeal in *Salomon* itself, one might have expected a more explicit discussion of the causes of this reversal in judicial policy. Instead, the change seems to have taken effect without much explanation as to why. This was a development that was described by Professor Kahn-Freund as being a 'calamitous decision' (*Kahn-Freund* (1944)). If that were so, why would it have been effected? Given the lack of explanation by the House of Lords we can only speculate on the underlying motives. My suggestion is that the late Victorian British Empire needed capital to finance trade and to finance the Empire itself. Entrepreneurs and groups of traders could be encouraged into activity far more easily if the company was a distinct legal person that could exist separately from its human parents, and therefore I have always assumed that the urge to develop the company on a model which would assist the economy was considered to be more important than the early 19th-century morality which balked at such things.

An alternative analysis (alternative, that is, from any of the judges in the House of Lords in the *Salomon* litigation) would have been to say that the company is a cypher for the personality of Salomon and therefore not at all a distinct person. This perception of companies as merely avatars behind which real people carried on their activities has a provenance in the common law. In a decision of Lord Pollexfen CJ in *Sandy's Case* in 1684 ((1684) Cobbett, *State Trials*, X, 371), his Lordship had criticised the joint stock company as being the:

> invisible merchant that no one knows where to find . . . [which] in judgment of law has neither soul nor conscience and yet forsooth are traders.

This language of 'conscience' is of course traditional equitable language and part of the law of trusts. However, between the 17th century and the late 19th century there had been a profound judicial reconsideration of the desirability of the company. Whereas companies had once been outlawed in the wake of the South Sea Bubble, by 1897 companies were seen as being persons and therefore occupying a lifeworld of their own outwith the lifeworlds of their members, employees and creditors. The decision in *Salomon* was built on a logic which was itself built on an ideology of the company as a sentient, if artificial, economic actor.

Lifting the corporate veil

The problem of lifting the corporate veil

What is important in the study of company law is that the English courts have been very reluctant indeed to look behind 'the corporate veil' – the idea that a company is a distinct legal person – except in a very limited category of circumstances. This section considers the cases on what is known as 'lifting the veil' of incorporation, which is one of the early components of most university company law courses. The expression 'lifting the veil' means ignoring the fact that an act has been performed by a company, which is a distinct legal person, and instead holding that any human beings involved should be personally liable as though there had not been a company there. Contrariwise, refusing to lift the veil means persisting with the legal principle that the company should be treated as a distinct legal person so that any human beings involved will not be held liable for any act done by the company. An excellent article by Ottolenghi (1990) attempts to unscramble the various approaches in the cases in this area by differentiating between cases in which the veil should be ignored (especially where there is fraud) and cases where there should be a limited lifting of the veil (for example to ensure equity in tax law) and the vast bulk of cases where no lifting takes place.

In essence, unsatisfactorily enough in practice, the principle is that the veil will not be lifted, except in the few cases where it has been.

The general principle against lifting the veil

Illustrations of the distinction between the company and shareholders in 'one-man companies'

The first category of exceptional cases relates to so-called 'one man companies' in which the company is created to act as a cypher for one individual. So, in *Lee v Lee's Air Farming Ltd* (1961), Lee had formed the company through which he carried on a business of air services for farms. He was the governing shareholder and controlling director and therefore administered the company entirely on his own. He was killed while flying one of the company's aircraft. His widow claimed that she was entitled to compensation from the company under the Workers' Compensation Act on the basis that her husband was a 'worker'. It was argued on behalf of those who took over the company latterly that Lee had not been a 'worker' because he had in fact run the company. However, it was held that Lee was a separate person from the company, Lee's Air Farming Ltd, and therefore there was no prohibition on claiming compensation from the company under the legislation as a 'worker' distinct from the employer company.

Similarly in *Macaura v Northern Assurance* (1925), it was held that the individual who controlled a company was not entitled to take out an insurance contract over property that belonged to the company that he had created. It is a key part of English insurance law that to insure property you must have an 'insurable interest', which means some form of ownership, in relation to that property. The individual had purported to take out an insurance policy over the company's property, which was subsequently destroyed. It was held, however, that the company was a distinct person and therefore it was only the company that could have an insurable interest in the company's property. The claimant could not have such an interest even if he was the sole human being involved in the ownership and management of the company. Consequently, it was held that the individual could not claim under the insurance contract because the company was a distinct legal person and thus the owner of its property.

In *Salomon* and in *Lee* we see the positive side of the principle that the veil of incorporation will enable individuals to avoid being responsible for the company's liabilities and to claim rights distinct from the companies that they control, even though the company is simply a tool created for their convenience. In *Macaura*, however, we see how an individual can lapse into the practice of assuming incorrectly that property owned by 'his' company is actually 'his own' property. Here we learn the valuable lesson that while the doctrine of corporate personality can be useful for business people, company law nevertheless requires that the individuals involved in a company must obey the legal niceties to get the most out of the company model.

The mainstream principle: that the veil will very rarely be lifted

Occasionally a case is heard in which the court decides to lift the veil of incorporation so as to treat the shareholders as being responsible for the acts or obligations of the company. So, in *Creasey v Breachwood Motors* (1993) the general manager of the Saab car dealership in Welwyn was summarily dismissed. At that time the Saab dealership had been operated by Welwyn Ltd, but latterly the business was transferred to Motors Ltd. It was suggested that Welwyn Ltd had been insolvent even though it had £70,000 in assets and only one creditor, with the result that assets were transferred between the two companies. The general manager sought to proceed against Motors Ltd on the basis that Welwyn Ltd had no assets left, and on the basis that Motors Ltd was effectively the same company as Welwyn Ltd carrying on the same business and being operated in practice by the same individuals. There was a suggestion in the judgment that there had perhaps been some skulduggery in the transfer of assets between companies, although that is not made entirely clear. It was held that this situation merited lifting the veil of incorporation so as to ignore the fact that there were two different companies involved.

This decision was overruled by the Court of Appeal in *Ord v Belhaven* (1998), where two individuals agreed to take a lease over a pub but alleged that they had been the victims of misrepresentation made on behalf of a defendant company as to the pub's profitability. The group of companies that contained the defendant company was reorganised. The company law issue was whether or not the claimants were restricted to proceeding against the company on whose behalf the misrepresentation had been made or whether the veil of incorporation could be lifted so that the claimants could proceed against a different company. It was held by Hobhouse LJ, reasserting the orthodox position in company law, that *Creasey* had been wrongly decided. There is a distinction, it is suggested, on the facts in that in *Ord* the transfer had taken effect at a valid book value and so it did not appear that there was any skulduggery involved. In essence, the claimants had no action because the company against which company law required them to proceed had no assets left. What should be noted, however, is that the veil of incorporation is only very rarely lifted.

Another way of thinking about the corporate veil: lifting the veil in cases of unconscionability and fraud

Among the cases in which the veil of incorporation has been lifted are cases in which a company is created solely to defraud the claimant or to elude the terms of a contract (see in particular *Woolfson v Strathclyde RC* (1978)). So, in *Gilford Motor Co Ltd v Horne* (1933), Horne signed a contract by which he agreed not to solicit the claimant's clients. Rather than solicit business himself, Horne created a company which he used to set up a business in direct competition with the claimant. Horne hoped that he would be able to argue that it was not he who had breached the contract, but rather that it was the company that had done it. It was held that the creation of this company was a mere facade and a sham to elude Horne's contractual obligations and therefore that the veil should be lifted and the claimant allowed to proceed against Horne. The deliberate breach of the contract justified lifting the veil of incorporation: that is, ignoring the fact that technically it was the company that had breached the contract, and instead holding Horne liable personally.

In *Jones v Lipman* (1962), the defendant had contracted to sell land to the claimant but later decided to try to elude that contract. To do so, he created a company and transferred the land to the company in the hope that he could argue that the land belonged to the company and that the company was not bound by the contract. It was held again that this company was being used solely as a sham so as to breach the defendant's contract. What these two cases have in common is the creation of the company solely to breach the defendant's own obligations after those obligations had come into existence, whereas at least A Salomon & Co Ltd was created to operate the business and not simply to elude a specific contractual obligation.

In Chapter 1 we considered how the Victorian company was often perceived, even in popular culture, as a means of effecting fraud. In *Re Darby* (1911) two well-known fraudsters, Darby and Glyde, promoted a company that claimed to have capital of £100,000 in issue when in fact only £11,000 had actually been issued. Investors acquired shares in this business in part because it seemed to be so well capitalised. The shysters Darby and Glyde then sold property to the company, effectively sucking all of the investment capital that they had raised from other people out of the company. Darby argued that he was a different person from the company and therefore that the liability for the lost investments lay with the company. It was held that the company here was clearly an alias, which was constituted to enable the two shysters to commit this fraud, and therefore the veil of incorporation was lifted so that Darby could be sued personally. This was another clear example of the veil of incorporation being lifted to prevent straightforward fraud.

Similarly in *Aveling Barford v Perion* (1989) a so-called asset-stripper sought to defraud a company that he had taken over by selling the company's assets to himself at an under-value, so that he could later sell them at their market value and keep the difference for himself. He argued that the company was the only possible defendant to any action to recover the loss on the transaction. It was held that this was a fraud and that the veil of incorporation should be lifted so that the asset-stripper could be sued personally. (In any event, for a fiduciary to deal with himself in this fashion would give rise to a constructive trust over those profits, as discussed in Chapter 7.)

My personal view, for what it is worth, is that the law in this area should be organised around an equitable principle that the veil of incorporation should be lifted if it can be shown that the defendant companies or the individuals who operate them have taken unconscionable advantage of the claimant. This would be based on the equitable principle that 'statute may not be used as an engine of fraud'. In essence, if reliance on the companies legislation, and also the common law rule that companies are separate people, would have the effect that the defendant would be defrauding the claimant or knowingly treating her unconscionably, then the veil of incorporation would be lifted. So, in *Creasey*, if it could be demonstrated that assets were transferred between companies so as to put assets unconscionably beyond the reach of the claimant, then that would permit the veil to be lifted. However, if there were some other, genuine commercial reason for transferring assets between companies, that would entitle the defendants to rely on the different personality of the various companies involved. I return to this idea below, but first let us consider a more recent case, which again considered a fraudulent scheme that used a company as a cypher for the individual fraudster.

Where the company is merely the alter ego of an individual: Stone Rolls v Moore Stephens

The decision of the House of Lords in *Stone Rolls Ltd v Moore Stephens* (2009) throws further light on the corporate personality of the company. In particular, it makes us question the nature of the interrelationship between the company and its directors. Given that a company is a mere device through which commerce and other activities can be carried on, how do we explain the activities of the human beings who perform acts under the cloak of corporate personality in legal terms? In short, any human being acting on behalf of a company is an agent of that company. But in what circumstances will it be appropriate to think of the company as being either the agent or the alter ego of the human beings who have brought it into existence in the first place?

In *Stone Rolls Ltd v Moore Stephens*, Mr Stojevic was accepted as being the 'very ego' of the company. The shares in that company were held by a Stojevic family trust, which concealed Mr Stojevic's de facto ownership and control of the company's shares through the trust. Moore Stephens were auditors, whose job as auditor was, simply put, to investigate the accuracy of the company's accounts (as discussed in Chapter 6). Mr Stojevic caused the company to commit a series of complex frauds that the auditor did not detect. In essence, Mr Stojevic conspired with others to have letters of credit issued in favour of the company, and then acquired loans fraudulently from banks based on those letters of credit. In time, when the frauds were uncovered, it was also discovered the company had suffered large losses because a large amount of money had been paid to the conspirators to the fraud. The company then sued the auditor for failing to uncover the frauds and thus protect the company from the ensuing losses.

Lord Phillips began his judgment with the robust statement that 'Mr Stojevic is a fraudster'. With typical clear-sightedness, his Lordship cut to the heart of the matter. And yet it was being claimed that the company should be able to recover damages from the auditors even though the root cause of the loss was the fraud committed by the human being who had established the company. Therefore, it was important to consider whether the company should be considered to be a mere alter ego of Stojevic, because if it was, then it could have been said that Stojevic would effectively have been recovering the loss which he had caused from a third party. Alternatively, if one sticks to the view that the company is a separate person, then the company should have been entitled to recover those losses.

It was held that the doctrine of *ex turpi causa* meant that a person could not rely on its own wrong to demonstrate a loss. On these facts, it was held that the company should be treated as being the alter ego of Mr Stojevic, and therefore the company was in effect relying on its own wrong in committing these frauds to prove the losses that it had suffered. It is suggested that if the court had been able to identify that the activities of the company and of Mr Stojevic had been unconscionable on the basis that they had been used to perpetrate a fraud, then that would also have justified lifting the veil of incorporation.

The dissenting judgments of Lord Scott and Lord Mance took the traditional company law view that Mr Stojevic was a person distinct from the company and therefore that the company should have been able to sue him personally for any wrong he had caused. If that was the case, the company was entitled to sue its auditors as a person distinct from the true fraudster, Mr Stojevic. This dissenting position, and the traditional company law position which it embodies, requires us to overlook the fact that the company existed to do the bidding of Mr Stojevic. There is, nevertheless, a question as to whether or not auditors should be responsible for losses which are caused by its actions to companies more generally: that question is considered in Chapter 6. It is complicated here because the only real victims of this fraud were the banks that had lent money to the company, because this company was merely the alter ego of Mr Stojevic, whereas in a normal trading company there would have been other shareholders in whose interests it would have been appropriate to allow the company to sue the auditors.

Further illustrations of the veil principle where individuals act through the company

The English courts' determination to keep the veil drawn means that even if individuals act on behalf of the company, the liability remains that of the company. So, in the House of Lords in *Williams Natural Life* (1998), it was held that the individual who ran the Rugby branch of a health food store through a company had not assumed liability for the acts of the company and therefore would not be personally liable for the obligations of the company, even though the claim was based on misrepresentations made in a brochure as to the company's profitability that the human defendant had made when acting on behalf of the company. Consequently, where a contract is made with a company, the managing director of that company will not be liable for misrepresentations made on behalf of that company unless he had expressly assumed responsibility for the company's liability as part of that same contract.

Similarly, in *Yukong v Rendsburg* (1998), a charterparty between Yukong and RIC, a company, was repudiated by RIC. RIC transferred its assets to another company, LIC, so that Yukong would not be able to have recourse to those assets when it sued RIC. Of course, the physical transfer of assets was not performed by RIC because RIC was a company: in the real world those assets were transferred by the defendant director acting on behalf of RIC. Yukong proceeded against that director personally, but it was held that the director was not personally liable for the defaults of the company, and thus that the veil should not be lifted.

From the perspective of the claimant in both of these cases, what has happened is that movement of assets between artificial companies by individuals has meant that the claimant could not effectively proceed against the company (because the companies had insufficient assets to meet the claim) and the claimant could not proceed against the individuals who had procured those movements of assets because the companies were the proper defendants. It is suggested that this is a clear example of the companies legislation being used as an engine of fraud, because the human beings are being shielded by the use of companies that in truth have no real assets. Consequently, a shyster could shield her own liability by creating a company with very little capital and acting through that company at all times. Seen differently, a lawyer could advise her client to insulate herself from liabilities in her business by acting through a company, which would absorb the blows in any litigation. A scoundrel can use a company as a sort of shield in these circumstances. One of the most disturbing examples of this principle at work arises in the next section.

Problems of morality: liability in groups of companies and in multinational companies

As a preface to this section it is useful to know that companies can be organised into groups, whereby one company (the 'holding company') owns all of the shares in 'subsidiary companies', and therefore that the shareholders in the holding company effectively hold all of the value in the subsidiary companies through the holding company. Nevertheless, each subsidiary company is considered by company law to be a distinct legal person, even though they have common ownership in the real world. Each subsidiary company in a group of companies must have its own directors, its own constitution, and so on, even though in practice the directors of the holding company and the subsidiary companies are often the same human beings. (The use of corporate groups is considered in detail in the next chapter.) It is also possible for groups of companies to be organised in different jurisdictions (known as a 'multinational company'): so a holding company in the UK can have subsidiary companies organised, for example, under the laws of South Africa and Australia to conduct businesses in South Africa and in Australia under the management ultimately of the UK executives. While the South African and Australian subsidiaries must comply with South African and Australian law respectively, they are considered by English law to be distinct legal entities from the UK holding company.

One of the most troubling decided cases on the *Salomon* principle, it seems to me, is that in *Adams v Cape Industries plc* (1990). The defendant company was part of a group of companies that operated mines in different countries. The holding company and the ultimate corporate management were located in the UK. In this instance, one of the subsidiary companies operated an asbestos mine in South Africa. Mineworkers developed very serious illnesses as a result of working in the South African subsidiary's mines and so brought a class action against that company for damages. However, the South African subsidiary had insufficient funds to meet the claim and so the miners tried to bring their claim against the holding company in the UK. (The holding company can be thought of here as being the head company with most of the group's assets and the senior management of the business.) It was held that the claimants could only bring a claim against the South African subsidiary with which they had contracts of employment, and that the UK holding company was a different legal person, which was not liable for the acts or omissions of the South African subsidiary company. Because those companies were considered to be separate legal persons, in spite of having common ownership and management, the liabilities of one company could not be imposed on another company.

There was an argument raised in that case as to whether or not the group was in truth a single economic unit and the group structure a mere facade, so that the veil of incorporation could be lifted. Nevertheless, it was held that this sort of organisation and the treatment of each subsidiary as a distinct legal person is 'inherent in our corporate law'. This decision followed the earlier decision in *The Albazero* (1977) in which it was held by Roskill J to be 'unchallengeable by judicial decision . . . that each company in a group [is a] separate legal entity'. The decision of the Court of Appeal in *DHN Food Distributors v Tower Hamlets* (1976) had nevertheless held that a holding company was entitled to recover damages in relation to disturbance caused to land which was held in the name of a subsidiary company because the holding company and the subsidiary in that instance constituted a single economic unit. In that case, the subsidiary was found to be merely holding the land on behalf of its holding company, and the holding company was found to have complete control of the subsidiary. That does not distinguish it from the normal situation in which subsidiary companies are usually controlled by their holding companies. This simply illustrates that in this area of law we can identify general principles, but then that we must acknowledge that on particular sets of facts a court may be convinced to lift the veil in spite of that general principle.

Nevertheless, we can learn a simple lesson. To restrict the possibility of being sued by third parties or employees, a business should divide its operations among a number of subsidiary companies, because the defaults of one company will not affect the other companies in the group, even though they are ultimately owned by the same people. The effect for very sick, impoverished African mineworkers was that a wealthy UK company, its shareholders and directors avoided liability for the defaults of things done and not done in the name of its South African subsidiary. It has also been held that subsidiaries will not be responsible for the defaults of the holding company in a group of companies (*Atlas Maritime v Avalon* (1991)). This is the troubling aspect of the *Salomon* principle: it can shield clever capitalists from liability and prevent other people from recovering what they might otherwise be owed. The moral questions that are raised by company law being used to avoid liability, for example, for serious personal injuries to employees are considered in Chapter 14 *Corporate social responsibility*.

My suggestion for the future of the veil principle

This is just my view about the veil principle and so you can ignore it if you like. The idea that a company is a distinct legal person is both deeply ingrained in our commercial life and clearly has a number of advantages in terms of business efficiency. Therefore, it is suggested that for nearly all practical purposes, the *Salomon* principle should be retained to simplify contracts and grease the wheels of commerce. However, when there is a suggestion that a person has sought to resist their personal obligations by hiding behind the corporate veil, then the court should ask itself whether or not the defendant is seeking to rely on the corporate veil unconscionably.

In essence, it is suggested that all of the cases on the corporate veil could be explained on the basis that the court considered that the defendant had or had not acted unconscionably. Consequently, where a person seeks to perpetrate fraud using the company, that would be unconscionable behaviour; where a person seeks to avoid a personal undertaking to the claimant on grounds that the company owed the obligation, the question should turn on whether or not the defendant was doing so unconscionably. It should become an equitable doctrine on the basis that, as with all equitable doctrines at root (from Aristotle onwards), the common law recognises the distinct personality of the company, and equity should consider whether or nor that strict common law rule has permitted unconscionable advantage to be taken of the claimant in any particular circumstance.

This is the basis of the law of trusts, on which company law was originally based, in which trustees are required to act conscionably, just as directors (who are also subject to fiduciary duties) should also be required to act. By way of illustration, *Lee* and *Macaura* did not involve taking advantage of the claimant, whereas *Gilford Motors* and *Jones v Lipman* did. Therefore, only the latter cases should involve lifting the veil of incorporation. My own personal view is that the use of the principle in *Adams v Cape Industries* constituted an unconscionable denial of human liability for harm caused to other people and therefore the veil should have been lifted on that equitable basis.

Wrongful and fraudulent trading

There are three statutory exceptions to the *Salomon* principle. The first relates to 'fraudulent trading' further to s 213 of the Insolvency Act 1986:

> (1) If in the course of the winding up of a company it appears that any business of the company has been carried on with intent to defraud creditors of the company . . .

(2) The court . . . may declare that any persons who are knowingly parties to the carrying on of the business . . . are liable to make such contributions . . . to the company's assets as the court thinks proper.

The 'winding up' of a company refers to the process of terminating it. If the termination of the company has been conducted so as to defraud the creditors – for example, by intentionally taking money from creditors and then seeking to wind the company up with the shield of limited liability – then anyone who is knowingly a party to the company's business is personally liable to make any contribution to the company's assets which the court decides to order. Therefore, individuals will be responsible for the company's obligations to its creditors in these circumstances even though in ordinary conditions the company would be solely responsible for its own obligations. It was held in *Re Patrick & Lyon Ltd* (1933) that in this context the term 'fraud' means 'actual dishonesty' which, 'according to current notions of fair trading among commercial men' constitutes 'real moral blame'.

The insolvency legislation provides for a further exception to the strict distinction that is ordinarily kept between the company and its human operators. The provisions dealing with wrongful trading are set out in s 214(2)(b) of the Insolvency Act 1986 to the effect that the same liability applies if 'that person knew or ought to have concluded that there was no reasonable prospect that the company would avoid going into insolvent liquidation'. It was held in *Re Produce Marketing Consortium Ltd (No. 2)* (1989) that much depends on the nature of company in question and whether in the context the directors should realise that the company is nearing insolvency.

Section 213 is a coercive provision and not simply a provision aimed at recovering damages. So in *Re a Company* (1991), it was held that this provision is intended to exact punishment and not simply to extract compensation. The punitive aspect of this area of law is expanded by s 993 of the Companies Act 2006 to the following effect:

If any business of a company is carried on with intent to defraud customers of the company . . . every person who is knowingly a party to the carrying on of the business in that manner commits an offence.

Therefore, if the business is carried on generally so as to defraud the 'customers' of the company, then anyone who knowingly participates in that fraud commits a criminal offence.

One member companies

The notion that a company is a person distinct from the human beings who operate it is generally comprehensible in the context of companies with a number of shareholders and a number of directors: that the company is a distinct legal person is useful because it means that the company can look after the best interests of all of the people who have a stake in the company. Where this idea stretches credulity would be where there is only one human being acting as shareholder and director, because the company would then simply be a cypher of that person.

It is possible since 2006 to have a private company with only one shareholder and one director: so-called 'one member companies', because there is only one member of the company (s 154 Companies Act 2006). Such a company would simply be acting as a cypher for that individual human being. There is no longer the requirement, as in the time of *Salomon v A Salomon & Co Ltd* (1897), to have more than one human being involved in the company. This intensifies the notion that the company's distinct personality is an artificial device. Importantly, this is not possible with public companies, and therefore to reach the same effect with a public company there would need to be other people installed as directors. The sort of difficulties and abuses which can be caused by having companies that are in fact controlled by one individual was illustrated by *Stone Rolls v Moore Stephens* as considered above. This is an acceptance of the sort of sleight of hand that was illustrated by the *Salomon* case itself being legitimised now by the companies legislation. Indeed, there is much about capitalism which has developed from the world of Aron Salomon: not least the shoe business, as we shall discover.

Corporate responsibility

In Chapter 4 we consider situations in which a company will be responsible for acts committed by human beings on its behalf for which the company will either be vicariously responsible or for which the acts of those people will be deemed to be the acts of the company. Those cases could be considered as being exceptions to the *Salomon* principle, although they are the inverse of most of the cases considered here, because here it has generally been a human being seeking to distance herself from liability for the obligations of the company on the basis that the company is a separate person. In truth, the corporate responsibility cases are slightly different because the existence of the company as a separate person is not being questioned, but the question is rather whether or not the directing mind and will of a company is acting *as the company* in particular circumstances so that the company should bear responsibility for what was done in its name by human beings. Nevertheless, those cases will tend to illustrate how in many circumstances it is necessary to blur the boundary between the company and the individual.

The effects of the *Salomon* principle today

The benefits of the principle

The main benefit that flows from the *Salomon* principle is one of efficiency. Whereas previously a business organised as a partnership could only create contracts in a very complicated way – involving each partner becoming a party to that contract, and involving the hidden nature of the rights and obligations of individual partners – as soon as it is recognised that a company is a distinct, legal person in itself, then the company can create contracts in its own name. As a result, the process of creating contracts with businesses became much simpler. The parties needed only to create one single contract with a human being who was authorised to create that contract on behalf of the company. (This leads some theorists, like Coase (1937), to refer to a company as being 'a nexus of contracts' in which a company gathers a number of contracts into one.) In practice, in relation to significant contracts, it is usual to have the people authorised to create contracts either identified in the articles of association of the company or agreed by a formal decree of the board of directors. The corporate personality of companies therefore makes trade simpler when it involves complex commercial organisations.

The second benefit is a more personal one from the perspective of those human beings who are inside the company. For an entrepreneur (or sole trader) like Aron Salomon, the benefit is that his own personal property can be protected from the failure of the business. If an entrepreneur dealt as a sole trader, then she would face the risk that all of her own property (her money, her house, her car, and so on) would be lost if the business failed. By organising her business as a company, the entrepreneur puts distance between her personal property and the company's property, so that only the company's property will be lost if the business fails.

The problems with the principle

The central problem with the *Salomon* principle is an ethical one. It is the inverse of the second benefit, discussed immediately above, when seen from the perspective of people dealing with the company from the outside. If Aron Salomon's property is protected, then people dealing with the company only have the company's own assets available to them if the company goes into insolvency. This means that an entrepreneur in the position of Aron Salomon may give less care and attention to the need to deal honestly and fairly with third parties because the entrepreneur faces no great personal risk of loss, beyond wounded pride, a tarnished reputation and the lost hope of a profitable business (but see what is said below about fraudulent trading). Similarly, other shareholders in a company bear no personal risk of loss if the company fails because the limited liability that is granted by our company law by definition limits their personal liabilities. When we add all of this together, we arrive at a position whereby the entire economy is peopled by companies whose shareholders and management bear little direct personal responsibility or loss if those companies should fail. The ethics of that economy become questionable if no one faces the risk of open-ended, personal loss.

The risks that shareholders and entrepreneurs do face

Of course, shareholders do face the risk of losing their investments in a company. The entrepreneur faces loss of reputation with investors and other traders. What they do not face is the loss of their personal property. There are criminal liabilities for fraudulent and wrongful trading, as discussed above, and so the human beings cannot simply use the company as a shield for any criminal activity, nor for any deliberate attempt to defraud third parties.

A moral distance between the company and the shareholders

As Professor Roger Cotterrell has pointed out, a company occupies a different moral position from the individual in that stigma attaching to the company for its actions will not necessarily translate directly into stigma attaching to any individual shareholder or director (Cotterrell 1992, p. 126). While, for example, the possibility of companies committing crimes has been accepted by English law, it is not such an obvious link to attach stigma for criminal or more generally 'immoral' activities to either directors, employees or shareholders of companies. The company-as-cypher enables individuals to hide behind the facade of corporate personality. This was exactly what Aron Salomon sought to do at the micro-level in the 1890s. As a result of his litigation, the same trick can be performed today by anyone who wants to invest in the arms industry or sweatshop factories without tarnishing their own personal images: liability and moral blame are restricted to the corporate cypher. This is true also of those who seek to avoid personal taxation by vesting their assets in offshore companies. What is different between 1897 and today is the scale on which it is possible for multinational companies to avoid or limit liability for their human agents. In Chapter 4 we shall consider the circumstances in which companies may nevertheless be held liable for the acts that are carried out by its human agents on its behalf, which is a slightly different question from the company being a distinct legal person, because it relates to the attribution of responsibility and vicarious liability, and not distinct personality.

This artificiality is, however, just part of corporate activity today. For example, in *Holland v Commissioners for Her Majesty's Revenue and Customs* (2010) the Supreme Court considered the respondent's business which offered 'off the shelf', shell companies to each of his clients to help them organise their tax affairs. As a result, a very large number of companies were created by the respondent to conduct this business which were just the tools of the respondent. Nevertheless, it was held that each company constituted a legal person distinct from the respondent himself, no matter how complex and artificial the arrangements were. That each company had separate personality was the basis of the respondent's business and so the court simply upheld the traditional principle.

A postscript to the Salomon case

As a postscript to the *Salomon* case, Rubin (1983) has unearthed some fascinating information about other aspects of Aron Salomon's business. Aron had brothers and friends, including Mr Claypole, with whom he set up other companies. The *Shoe and Leather Record* (a trade newspaper at the time) even went so far as to ask whether there was any difference between A Salomon & Co Ltd and other companies, Fitzwell Ltd and Claypole Ltd, which were a number of businesses in the boot trade under the ownership of the same gaggle of human beings. The suggestion seems to have been that these companies were operating as cyphers for those human beings in a way that would have seemed very unusual at the time. A number of other cases sprang from the operation of these other companies (for example, *In re Raphael* (1899); *Salomon & Co v Claypole Ltd* (unreported)), although none as celebrated as the decision of the House of Lords which was discussed above. The suggestion is that this group of family and friends were clever organisers of their businesses and of their personal affairs, even if the economic climate drove those businesses into failure.

As to the popular conception of the morality of the *Salomon* case in the general press, the *Shoe and Leather Record* considered that, even though 'it is not necessary to impute evil motives to Mr Salomon' (in spite of having done so inferentially by mentioning the idea), 'the result of these transactions has been to put a considerable sum into the pocket of Mr Salomon in his personal capacity which properly belongs to the creditors of the same gentleman in his corporate capacity'. So, it was suggested in the trade press at the time that Aron Salomon should have given up the great wealth he had amassed for himself and instead paid it out to his creditors to whom it was said to have belonged morally. An article in the *Law Quarterly Review* ((1899) 15 LQR 235) referred to Aron Salomon as being a 'ruined gamester' or gambler. With all of this opprobrium directed at a prominent Jew who supported a number of Jewish charities in London's East End and who had arrived in London in his twenties from Germany, one cannot but wonder whether there was the slightest whiff of anti-Semitism in some of this commentary or whether it was simply a distaste for the sort of sharp commercial practice which would become commonplace in the 20th century.

The **Salomon** *legacy and my running shoes*

It just so happens that I own a number of pairs of trainers that bear a 'Salomon' brand. The logo is two semicircles entwined so as to form a stylised letter 'S'. Aron Salomon manufactured boots in the East End of London in the 19th century and his plight gave rise to the birth of one of the foundational principles of modern company law; in the 21st century, coincidentally, there is a company with the same name which manufactures walking boots and trail running shoes. (I once terrified a lecture hall of students by tearing one of these trainers off my foot (having worn it specially for the occasion) and hefting it onto the projector at the front of the room for all to see.)

As befits the 21st century, modern Salomon footwear is sold over the internet as well as in shops. Their footwear naturally bears the corporate logo, and also a number of trade marks, runes and names which play heavily on the technological advances which they are said to embody ('Gore-Tex fabric', flex technology, and '3-D' soles designed to cope three-dimensionally with rough terrain – although, how a shoe could exist in less than three dimensions is not entirely clear to me). Their adverts present a clear brand image of people running healthily and joyously up sunlit mountains. (If like me you are in your early forties and so are constantly trying to grasp at the last of your youth and good health, you tend to look at advertisements showing fit people running up mountainsides with a pang.)

Companies have been legal persons ever since 1897, but today many companies have another type of personality as well: they have an image, a personality (in the ordinary sense of that term) and a reputation. Companies' brands tell of technological advances, reliability, energy, fun, reassurance, thrills, or whatever else they promise to bring into our shallow little lives. The seed planted by the *Salomon* case continues to mature in a way that embodies a virtual life for our companies as tools of businesses that want to market themselves to other people, real and virtual. The Salomon trainers, which were manufactured (worryingly) in the Far East (as I discovered when they arrived through the post) and in which I run irregularly round a lake in rural Sussex, are a sign that that seed has grown into a mighty tree. We shall consider the Salomon brand in the final chapter, but for now we must consider in the next chapter how modern companies are structured before coming to the legal rules governing their operation in practice.

Chapter 3

The nature of modern companies

Introduction

This chapter considers the various types of modern company. Modern companies are used for many more purposes than were ever considered by the people who drafted the early companies legislation. Just as Aron Salomon hit upon an imaginative use of the limited liability company (as considered in the previous chapter), so there have been many more uses developed by clever practitioners to help their clients since then. Aron Salomon used the company to avoid the possibility of personal liability for the potential failure of his business. However, as we shall see, most companies in existence today have no trade at all because they are used either simply to hold assets or as holding companies in trading groups. Most company law scholarship is concerned with trading companies, although there are many differences between individual trading companies. We shall consider the different ways in which various types of people may think about the same trading company: whether as creditors, employees, directors, customers, policymakers, or long-term or short-term shareholders. We shall also consider how different types of company have different impacts on the economy. Finally, we shall also consider the various ways in which modern company law theory has explained the nature of the company. Our consideration of company law in this book will be greatly assisted by understanding how companies are used in the real world and how they affect ordinary people.

The purposes and uses of companies

Not all companies carry on business activities: asset management and regulatory avoidance

There is an enthusiastic prejudice among those who write about company law and in the company law jurisprudence that all companies carry on a trade and that the hero of the company law story is the *entrepreneur*. Most good stories begin with the legend 'Once upon a time . . .'; most biographies begin with a person's birth and follow it through to death; and most accounts of company law begin with the creation of a company by an entrepreneur. An entrepreneur in popular legend is a businessperson who has a lantern jaw, a resolute purpose and a thousand-yard stare. Entrepreneurs create businesses out of thin air and provide jobs, prosperity and economic growth. Entrepreneurs are always presented as being the engines of the economy. When entrepreneurs set up their life-giving commercial vehicles, they are said to choose the company as the form for their business because it offers them limited liability (in case the business should fail), ease of contracting, and all of the other advantages considered at the end of the last chapter. However, companies already exist in the world as we talk. So we cannot think of all companies as being entrepreneurial vehicles that are brought into being from scratch by these mythical heroes of the economy; instead we must also think about the way in which the law treats companies that are already in existence.

In fact, most companies in the world conduct no trade at all, nor do they employ people, nor do they contribute directly to the economy. Rather, they are vehicles created to hold assets. We shall refer to these companies as being *asset management companies*. One of the advantages of the company is that it can be identified as the owner of property so that the human controller of that property can conceal her ownership or control of the company's property from the company's creditors or from regulators, or she can use it to minimise her liability to tax. If the company owns the property, then it is usual to expect that the company is responsible for all obligations connected to that property. If a human being wants to minimise her potential liabilities in relation to that property, then she could transfer the property to a company and hold all of the shares in that company herself. This sort of company may well be a *regulatory avoidance company* in that its purpose is to avoid regulatory oversight by shifting any regulatory penalties or costs onto the company and away from the individual. This is particularly useful if the company pays a lower rate of tax than the individual, or if the company is resident in another country beyond the reach of the regulator or the English courts.

Groups of companies

As is considered below, most corporate trading companies, as well as regulatory avoidance companies, are organised as groups: that is, the shares in a company are owned by another company (a 'holding company') instead of being owned by a human being. It is very common for companies to own shares in other companies. As a result it becomes very difficult to trace the ultimate beneficial ownership of companies in some instances. So, in a group of companies there will be a *holding company* that owns all of the shares in the *subsidiary companies* that comprise the rest of that group of companies. Each subsidiary company will carry on a particular business activity or hold identified assets, whereas the holding company is the central part of the structure through which senior management will ordinarily be carried on (like a spider at the centre of a web) and which commonly

only owns shares in subsidiary companies but no other property. The holding company will own all (or most) of the shares in the subsidiary companies.

An example may make this clearer. If an undertaking comprises one business that manufactures shoes and another business that operates shoe shops, it is likely that one subsidiary company would own and operate the shoe manufacturing business while a second subsidiary company would own and operate the shoe shop business, and the shares in both of those subsidiary companies would be owned in turn by a holding company through which senior management would control both businesses and the entire group of companies. One advantage of this structure would mean that it would be possible to sell off either business simply by selling the shares in the appropriate company if management wanted to concentrate on the other business. Another advantage would be that the structure of the entire organisation can be kept simple by separating off the ownership of different premises, the tax affairs of the organisation can be controlled efficiently, and so on. (A good example of a very complicated arrangement in which companies were organised solely to avoid liability to tax is set out in *ABC Ltd v M* (2002).)

There may be situations in which, for example, a multinational group of companies manufactures training shoes through a number of subsidiary companies in different jurisdictions where labour is cheap, but which has its holding company organised in the UK because senior management wishes to remain in the UK, while each different jurisdiction may have its own subsidiary company to simplify the structure of the business. It becomes easier to move production to whichever jurisdiction has the lowest wages simply by transferring activities between the different companies in the different countries. (More commonly today such multinational companies prefer to use a franchising model in which they contract with entirely separate manufacturers in other jurisdictions to make goods for them for a limited time period, and so hope to prevent any bad reputation or other liability attaching to any part of their group of companies.)

In banking organisations there are commonly many hundreds of subsidiary companies in different jurisdictions that carry on different trading activities separately from other parts of the bank. From the outside, it appears that the bank is one entity behind its logo, but in legal terms each subsidiary company is technically a separate legal person. Usually this complex sort of organisation is used for different business areas in banking because different types of banking business have different regulatory controls imposed on them by different regulators, which require the companies which they regulate to put aside different amounts of capital to cover potential future losses. It is common even to use a separate company to hire vending machines for the office building, and another one to buy computers for the business, and so on. In this way the various liabilities of various parts of the business are separated off by using a range of different subsidiary companies. Outsiders dealing with the bank, assuming they have sufficient bargaining power, will want to have guarantees or other credit protection provided to them so that obligations that are owed to them by one subsidiary company are nevertheless guaranteed to be paid by the holding company (or by other group companies with more assets and income than the specific subsidiary company with which the outsider has contracted).

The use of complex groups of companies can therefore be a convenient way of managing both simple and highly complex businesses. It also makes it easier to sell off an individual business unit if that business unit is owned entirely by a single subsidiary company, because all that needs to be done is that the shares in that subsidiary company are sold to the purchaser. So, from the seed that was sown by the *Salomon* litigation, we have developed very sophisticated means of organising and administering complex commercial activity.

The modern twist on an ancient idea

Many subsidiary companies do not conduct businesses but rather hold assets within a group of companies; and as mentioned above many companies exist solely as asset management vehicles. In fact, most companies in existence are there solely to hold assets and not to trade. The same general company law nevertheless governs these asset-holding and regulatory avoidance companies as governs trading companies. US President Barack Obama identified this problem when he spoke of a single office block in the Cayman Islands in which there were 12,000 registered companies. Clearly, 12,000 trading businesses could not have been operating out of one, modest-sized building. Of course, it is perfectly possible for 12,000 companies that exist only as records on a computerised register to have their headquarters nominally in the same building because they take up no physical space. Therefore, what President Obama was referring to was the creation of *front companies* by professional investment managers in those islands to conceal investment income or other property from regulatory oversight in the USA and other countries. A front company is a company that merely creates the technical legal fiction that a particular group of assets are owned by that particular company in a way that distances them from the human beings who operate that company. It is possible, for example, for rich UK taxpayers to lodge some of their income-generating assets with a Cayman Islands company created specifically for that purpose in an effort to avoid tax in the UK.

A company, being intangible and merely an idea, needs no physical space in which to act. This sort of company is simply a *legal fiction*. It only exists because it is recognised as existing by the law. Suppose a rich individual wanted to conceal her assets from the tax authorities in the UK. If Aron Salomon could deny all responsibility for his business's debts simply by registering a company and repainting the sign over the gate to his premises, then all that needs to be done to organise 'ownership' of assets in a jurisdiction in which no tax is payable (a 'tax haven') is to register a company in that jurisdiction and transfer assets to that company on paper. As a matter of tax law it would be argued that those valuable assets were now owned by a company resident in that tax haven. Such asset management companies are often bought 'off the shelf' from corporate service providers in those jurisdictions (in fact they are often subsidiaries of international banks or accountancy or law firms). Such companies are known as 'shelf companies', or 'shell companies'. They do not carry on a business: instead they are asset management companies that operate as mere shells to organise ownership of property. Therefore, an infinite number of companies can be registered as 'existing' in a single building because companies need exist only in electronic registers and in legal theory, without the need for any physical space. This is light years away from what the 19th-century legislators had in mind, but it is activity that is nevertheless built on the selfsame company structure that they created.

So let us not become dewy-eyed about the entrepreneur and the unsullied good that company law does for our economy. Instead, let us recognise the world as it is. There are, it is true, plenty of companies created by entrepreneurs to stimulate economic activity, but that does not account for all of the companies in the world. Company law offers a great deal of convenience in commercial transactions and it also offers the means of achieving other goals. As we observed in the Introduction to this book, there is nothing necessarily good or bad in the company model; rather, it all depends on how the company is used. So in the sections to follow we shall consider some of the different concepts of a company.

The structure of trading undertakings using companies

An example of the use of a company to operate a business

If companies are simply tools that can be fashioned into a number of different forms to pursue a number of different goals, then it might be useful to give a couple of examples of how individual companies and groups of companies might be used. Let us imagine a trading company that has been created by the Gilbert family to organise their farming business. (If you have listened to *The Archers* on BBC Radio 4 (willingly or inadvertently) then this context may be familiar to you.) Let us suppose then that the Gilbert family business is divided into three parts: a general farming business including a dairy herd, a few pigs and vegetable-growing; a small farm shop selling produce directly to the public; and a sideline in manufacturing sausages. Let us suppose further that all members of the Gilbert family are involved in these businesses.

As company lawyers we might well advise them to organise their businesses in the following way. The adult members of the Gilbert family would become the original shareholders of Gilbert Farms Ltd so that ownership and control of the farm is divided up among them in the shares they agree. Let us assume that each member of the family invests £50,000 from their own money and receives 50,000 shares with a nominal value of £1 each (that is, the value attributed to those shares at the date of issue). The benefit for the Gilberts is that even if the business fails, it is the company that bears those losses and not the Gilberts personally. What the Gilberts risk losing if the company goes into insolvency are their personal investments of £50,000, because if the company is insolvent, it is worthless. Importantly, however, they would not have to meet the company's obligations to its creditors. The creditors can only sue the company. So, the Gilberts would be protected to that extent. This is the result of the *Salomon* principle discussed in the previous chapter.

The risks borne by each element of the family business are, however, very different. As farmers, or people who listen to *The Archers*, will tell you, traditional farming is dependable at one level, and yet it is becoming harder and harder for small farms to make profits. Consequently, it is common for smallholders to branch out and try to run businesses which will attract consumers looking for local foods (as in farm shops) or which will appeal to the gourmet food market (as with speciality sausages). However, these new businesses could fail. Therefore, the family members might sensibly decide to divide the different businesses between different subsidiary companies (for example, Gilbert Farms Ltd to run the farm, Gilbert Farm Shop Ltd to operate the farm shop alone, and Gilbert Sausages Ltd to make and sell the sausages), so that if one business should fail, then the other businesses would be shielded from the debts and obligations of that particular business. Again, using the *Salomon* principle and separating the businesses off into different companies, the different companies are separate from one another and so the losses attributable to each company are ring-fenced from the other companies.

To complete the corporate structure, the Gilberts would be required to have a holding company (let us call it Gilbert Holdings Ltd), which would own all of the shares in the three subsidiary companies; and in turn the shares in Gilbert Holdings Ltd would be owned by the Gilbert family members in proportions they would agree between themselves. Any third party dealing with any of the Gilbert companies would be well advised to seek guarantees of payments of any amounts owed to it from all of the other companies and from the members of the Gilbert family personally, because otherwise the *Salomon* principle would insulate each of the businesses and the individuals involved from one another. If you are a creditor of any of the Gilbert businesses, you cannot know whether or not the Gilberts are actually concentrating all of the value of their farming activities in one particular company. Therefore, you would be well advised to ensure, even if your contract is with one particular company within the Gilbert group, that you have a guarantee of performance from all of the other companies in the group. As we can see, company law does not just offer structuring possibilities to the company's shareholders; instead it also requires knowledge, care and expertise on the part of outsiders dealing with companies.

The difference between private companies and public companies

Company law draws a distinction between public companies whose shares can be bought and sold by the general public on stock markets, and private companies whose shares are not available to the public at large (Companies Act 2006, s 755).

The public companies (identifiable by the letters 'plc' after their name) are often household names. In policy terms, the fact that their shares are available to the general public means that they are necessarily made subject to more intrusive regulatory regimes than private companies whose shares are not offered to the general public. Private companies (identifiable by the letters 'ltd' after their names) therefore tend to be smaller in cash terms, ordinarily they will not be household names, and their activities are ordinarily conducted on a much smaller scale. If the Gilbert farm business is a comparatively small undertaking, it will doubtless be organised as a private company, whereas if the business was a very large industrial concern requiring large amounts of investment capital, it is more likely that it would be organised as a public company. For the operators of a public company the principal attraction to holding that status is that they are able to access large capital markets such as the London Stock Exchange to raise capital, whereas if their business was organised as a private company they would not be able to do so. This issue is considered in the next section. The distinction between public companies and private companies is considered in greater detail in Chapter 6.

Trading as a public company

Importantly, private companies may not offer shares in their companies to the public (CA 2006, s 755). Therefore, the Gilberts would not be able to have the shares in their companies quoted on the London Stock Exchange if they do organise their business into private companies. The rules relating to public companies are different from the rules relating to private companies precisely because the general public may be involved in acquiring shares in public companies, and a large amount of information is therefore required by the securities regulations considered in Chapter 11 to be made available to the public by the company. It is possible to convert a public company into a private company (referred to as 'taking the company private') or to convert a private company into a public company (known colloquially as 'going public').

If the Gilberts' farm business was in fact an enormously successful food production business, then the Gilberts might decide to convert their businesses into a *public limited company* (a 'plc'), because having shares quoted on the London Stock Exchange has the advantage of increasing the number of potential investors in their business enormously. Many millions of pounds could be raised for the company by issuing shares to the public and marketing those shares appropriately. The downside of being a public limited company rests principally in the burden of complying with the securities regulations considered in Chapter 11. If the Gilbert business expanded into a very large 'agri-business' undertaking, which farmed a huge amount of land and supplied supermarket chains with large amounts of produce, then it might seem to the family members that to expand and consolidate their business, they should raise money by offering shares to the public. Most of the shares in public companies are owned by pension funds and other similarly enormous 'institutional investors', rather than by individual human beings. Therefore, the institutional investors become very important to public companies when seeking to raise capital for their activities. These investors are considered further below.

So what we have taken from the discussion so far is an understanding of the ways in which different corporate forms in different types of company can be used to achieve different objectives. Different types of business, different types of activity, require different types of company. As was discussed in the Introduction, the company is simply a mindless tool which its human operators can fashion into whatever form seems most useful to them. Just as Dr Frankenstein moulded his creation to achieve his own diabolical ends, the Gilberts will be able to form their own creations to achieve their infinitely more wholesome objectives. As we said at the outset, nothing is either good or bad, but rather thinking makes it so. We turn in the next section to consider the shareholders' perspective.

Different perspectives on the company's role

There are a number of different intellectual perspectives on the nature of a company. They arise outside companies law, but they are significant for us when thinking about how modern companies function, and who are the various stakeholders within a company. A few of those ideas are considered here.

The company as an investment

From a shareholder's perspective, a company is a means of making an investment (Ireland, 1999). A shareholder *invests* in a company when buying shares. All investment involves risk, and so a shareholder is taking a range of risks when investing in a company. As an investment, a share provides the shareholder with an asset that can be sold for a profit if the company is seen as having become more valuable (whether because it is more profitable or because its business sector is expected to become more profitable or its assets have increased in value). If the company's shares are quoted on the London Stock Exchange, there will be an accessible market for those shares, whereas if the share is in a private company, those shares cannot be sold on the Stock Exchange, with the result that selling them will be more difficult. Private companies, as their name suggests, therefore tend to have more intimate groups of shareholders. A shareholder might hold shares in the company because she is one of its managers or alternatively she may hold that share simply with a view to making profits in the short term from selling it privately. Alternatively, the shareholder may be hoping that the company will be sufficiently profitable in each year of trading so that there will be sufficient realisable profits to pay her a *dividend* (that is, a proportion of the profits in that year which the directors consider are capable of being distributed to the shareholders).

Therefore, there are at least two reasons for making an investment in a company. First, an investment may be made with the intention of generating regular income by way of dividend for a shareholder. The shareholder may be interested only in the amount of money that is made, without having any particular interest in the activities that the company performs. An example of this sort of investor would be a pension fund (an example of a large-scale institutional investor) that is concerned only to generate a steady income for its pensioners. Alternatively, you might invest in a company in which you have a genuine interest, either in the sense of wanting to support its activities or because you are personally involved in the business. That investment will acquire the shareholder a right to receive dividend income and also a right to vote at the company's annual general meetings.

Second, the investment may be made with a view to a short-term gain, which is not predicated on any particular affinity with the company but rather with a view to making a quick profit in the expectation that the price of that share will increase so that it can be sold for a profit. This sort of speculation is not as concerned with the intrinsic qualities of the company as it is with a guess as to whether or not the stock market will consider that share to be more valuable in the future than it is at the date the shareholder buys it. In general terms, the value of a company's shares on the London Stock Exchange is measured by reference to its 'earnings per share': that is, the amount of distributable profit which is earned relative to the number of shares which are in issue. There is therefore great pressure on public companies quoted on the London Stock Exchange to generate as high a profit as possible for each share in issue. It is feared that this can distort corporate activities by favouring short-term profit over long-term business growth. It also concentrates the attention of management on continuing to grow, rather than consolidating their business.

A nexus of contracts

Another explanation of the nature of a company is as a nexus of contracts. This is an approach which could be associated with the economist Ronald Coase (1937) who explained that the attraction of a 'firm' is that it reduces transaction costs by concentrating all of the various contracts which otherwise would need to be made between the human beings involved into a single transaction with a company.

Let me explain this idea about transaction costs a little more clearly. If you recall, as was discussed in Chapter 1, joint stock companies were formed as partnerships between the various traders who wanted to join together. Partnership law requires that there be a contract between the partners and that that contract express who is entitled to contract on behalf of the partnership, which typically meant that outsiders dealing with the partnership needed to enter into a contract with all of the partners. One of the benefits of the company model is that authorised agents of the company (usually directors and employees) can create contracts and bind the company without needing to obtain the consent of all of the shareholders. In that sense, a company with separate personality on the *Salomon* model itself stands for the previous web of contracts that were required between the partners and third parties, and so the advent of companies with separate legal personality has made transactions much cheaper than previously. It is clearly more convenient to be able to organise one's business affairs through the medium of a company, or as discussed above through a group of companies, in any one of a number of different ways.

Institutional shareholding and shareholder influence

Concentrating specifically on public companies, so-called institutional investors have become the most significant shareholders in companies quoted on the London Stock Exchange. In particular, pension funds and financial institutions operating investment funds have acquired very large proportions of the equity (a financier's term for shares) quoted on the London Stock Exchange and on other exchanges around the world, including online markets for securities and other trading platforms operated by financial institutions. The investment decisions taken by these institutions have a large impact not only on the share price of companies but also on the resulting ability of those companies to access funding in securities markets. If a company's share price falls, then that is an expression of the market's opinion that the company is a bad bet, and therefore banks and other financial situations will be less eager to lend it money or to buy its shares in the future except at a disadvantageous price.

Consequently, it is part of the role of senior management in such companies to maintain good relations with financial institutions by means of a flow of information to them. As a result, institutional investors will receive separate briefings from companies on their results, on their strategic vision and so forth, in parallel to the presentations made to the companies' bankers, to the shareholders in general meetings and occasionally to the media more generally. These briefings will be closed meetings, off-the-record and intended either to pacify or to entice the professional investment community with information about the company's prospects. As a result we might think that these institutional investors, even though they own the same ordinary shares as everyone else for the most part, have in fact become a different category of shareholder that has privileged access to company management simply because they own much larger blocks of shares than smaller investors. Consequently, there is a two-tier split in shareholders between the powerful institutional investors who have the ear of the company's managers and the mass of shareholders who receive nothing more than the legally prescribed information.

The company as a legal fiction

When I was an undergraduate, I read a book by Lon Fuller (1967) called *Legal Fictions*, which helped my understanding of law enormously. Professor Fuller explains how law requires fictions to work in many circumstances. His understanding of a 'fiction' is as something positive: an intellectual device that allows us to achieve a useful result. For example, at a simple level, when a person evidences an intention to benefit another person with some of her property, then English trusts law *deems* that an equitable proprietary right passes to the intended beneficiaries of the arrangement and that simultaneously legal title in that property vests in the intended trustees of the arrangement. None of these things happen in nature, nor does the owner of the property have to intend these detailed actions. Instead the law *deems* that they have happened so that, conveniently, the beneficiaries acquire rights against the trustees and the trustees simultaneously acquire the common law rights necessary to fulfil their fiduciary duties. Legal fictions in this sense oil the wheels of our jurisprudence.

Professor Fuller also traced the source of the word 'person' to the Latin word 'persona', which literally meant the mask that an actor wore in classical drama. (This idea also appears in Hobbes's *Leviathan* of 1651.) It does not mean the actor as a human being; rather, it means the mask that the actor wore to disguise himself as the character he was portraying. (To think of it another way, psychoanalysts like Sigmund Freud in *Civilisation and its Discontents* (1929) refer to the veneer of civilisation with which we cover up our baser natures as being a sort of mask: concealing the conflict between ego and id.) As Professor Fuller identified, the idea of a 'person' was therefore the idea of a mask or facade. A company is just such a facade: the law treats it as being a person even though it is simply a mask behind which human beings (the shareholders and directors) act anonymously. The French legal term for a company is a société anonyme (abbreviated as SA), which means literally an anonymous association. This always seems to me to be a lyrical description of a company: it is an anonymous mask behind which and through which human beings act. As we identified in Chapter 2, the company is a useful fiction in this regard.

Company law is based on a huge number of legal fictions because they are commercially useful. The principal fiction, of course, is the company itself. By recognising that a company has legal personality, we are involved in a fiction because the company has no tangible personality at all except for that accorded to it by company law. Nevertheless, the fiction is used because it has commercially convenient results: the company can enter into contracts in its own name, it can raise capital, and commerce can be conducted with greater ease as a result. Consequently, fictions are not necessarily bad things. They may enable us to achieve the results we want to achieve without causing too much distress to the logical underpinnings of our legal system.

The courts have *deemed* that the company has separate personality because that has convenient results, and in turn that has given rise to a number of unintended commercial uses. It should come as no surprise that when we start with one fiction, a number of other fictions must follow. Indeed, the whole of company law is simply made up. None of it exists in nature. That human beings walk on two legs and breathe air and can fall in love does not necessarily mean that there must be companies that have separate legal personality. Instead we choose to believe that

that is the case. Some of the most artificial ideas in company law relate to the creation and maintenance of the company's accounts and its capital, all of which are combinations of accounting fictions and sensible rules to prevent the company's money being bled out of it without anyone knowing. The need for detailed rules about the company's capital flows from the initial fiction that the company has legal personality. One supposition is built on top of another. The policy prerogative of preventing fraud and the devaluation of companies therefore requires that further rules, presumptions and fictions be created to maintain the viability of the first fiction. It will help us to understand company law if we bear all of this in mind: each principle is just another means of preventing unfortunate outcomes or to protect the logic of the company with separate personality.

Other business models

As was discussed in the previous chapter, the company is really just one of a number of models that businesspeople and others can use to achieve their goals. The differences between companies and other potential business models can be stated briefly. This is often a part of a company law course, but in truth the distinctions between the various forms of business entity are not difficult to state. The advantages of the company have been considered already: limited liability providing protection against the failure of the business, the ability to organise the business into distinct units, and the distinct personality of the business leading to simplified transactions are the principal ones. However, it will not emerge until later chapters how bureaucratic company law is, in addition to the onerous obligations that are placed on its directors not simply to please themselves (as a sole trader could do) but rather to consider the legal rights of others. It is possible to transfer a business into a company by 'incorporating' it: that is, by creating a company and transferring the assets to the company, appointing directors, allotting shares to the people who will be the shareholders and complying with the formal rigmarole considered in Chapters 4 through 6 and in Chapter 10. However, we shall consider some pen portraits of some of the most significant business models that are not companies.

Before companies began to dominate corporate life, partnerships were the most important model. Ordinary partnerships are based on contract in that the partners create a contract between themselves (usually referred to as the partnership agreement). A partnership – which is referred to by lawyers as a 'firm' – does not have separate legal personality. Instead it is a description of a contractual relationship between the partners. Furthermore, partners owe one another fiduciary duties. Under s 1 of the Partnership Act 1890, a partnership is an undertaking 'in common with a view of profit'. Therefore, in any situation in which people came together seeking to earn profits could be deemed to be a partnership under English law: this potential uncertainty about whether or not an informal arrangement would be deemed to be a partnership creating fiduciary responsibilities for all of the participants might be an unattractive feature to commercial people demanding certainty about their rights and responsibilities. There are two principal difficulties with a partnership. First, the partners are personally liable for all of the partnership's losses. This means that a partner will not have her personal liability limited, thus opening her up to the risk of losing her home and being made personally bankrupt if the business should fail. Second, contracts made on behalf of the partnership must be made by each partner separately because the partnership is not itself a legal entity (although many partnerships formally delegate the power to act for the other partners in certain circumstances). This sort of complexity in transacting is swept away by the use of a company with distinct legal personality. The company is much neater than an ordinary partnership in that the company is a distinct legal person, in that the company can issue shares to raise capital, in that the directors and shareholders do not need to become parties to the company's contracts, and so on.

The limited liability partnership (LLP) was created by the Limited Liability Partnership Act 2000, so that the LLP is a body corporate with distinct legal personality and so that the liabilities of its members are limited. The LLP does involve formalities such as the need to be registered with Companies House and to lodge an 'incorporation document'. Yet, the members of an LLP do not owe the duties of ordinary partners to one another. All of these facets make an LLP more attractive than an ordinary partnership (principally by reducing the personal liabilities of the members and by separating the members off from one another). The result is a halfway house between the traditional partnership and the company.

There is also another form of partnership, the 'limited partnership', which was formed by the Limited Partnership Act 1907. In essence, there must be at least one 'general partner' who is responsible for all of the losses of the firm; and there can also be 'limited partners' whose liabilities are limited to an amount specified in the partnership agreement. Such partnerships must be registered with the registrar of companies. A limited partnership has a series of maximum numbers of partners imposed on it. The purpose of the limited partnership was to mimic a French structure which both provides that there will be someone who is liable for all of the losses of a business, while allowing other investors to be brought into the partnership with only limited liability.

An individual going into business on his own, like Aron Salomon, could set up as a sole trader and be the outright owner of both the business's assets and all of its liabilities. This means that the entrepreneur does not have the protection of limited liability, and the entrepreneur will also be covered by a different tax code from small companies. A sole trader operates in her own name and is the absolute owner of all of the assets used in the business. The risk for the sole trader is that if the business fails, it is the sole trader personally who bears all of the losses – risking bankruptcy ultimately. On the positive side, a sole trader needs very little capital to start up in business and has none of the large administrative burdens placed on her that apply to companies (as discussed in Chapter 6). For a sole trader, drawing up accounts and making decisions about the business are comparatively straightforward matters, unlike the bureaucracy that is required by the Companies Act 2006 and its attendant regulations. Setting up in business as a plumber or as a jobbing builder is often easier as a sole trader in the first instance because you can focus on getting business and practising your trade, rather than spending too much time on paperwork. However, if the business starts to grow and the risks of loss associated with it become greater, then the protection of limited liability and the formalities involved with operating a company and meeting with other directors may become an advantage.

Moving on ...

Having considered some of the commercial questions associated with the establishment of a company, some of the different understandings of what a company is, and some other forms of legal entity, it is time to consider how companies are structured and organised in practice, and the parameters of corporate responsibility.

Chapter 4

The legal structure of a company

Introduction

This chapter sets out the key concepts relating to the legal structure of a company. It describes in legal terms how companies are created and how their constitutions are established. What this chapter will also do is to introduce you to some of the key personnel and some of the key terminology relating to companies in UK law, which is in essence an extended glossary defining key terms and explaining where those concepts are discussed in greater detail elsewhere in the book.

All statutory references in this chapter are to the Companies Act 2006.

The administrative nature of company law

Much of company law is concerned with administration. The Companies Act 2006 (CA 2006) is primarily a compilation of detailed rules as to registration requirements, constitutional requirements within the company's documentation, the powers and obligations of the company's officers in various contexts, the mechanism for making decisions within a company, and obligations to publish identified types of information at identified times. We shall be picking out some key provisions. Before we can come to those detailed provisions, however, we should fill in some of the basic terminology relating to companies.

The essentials of company law

A company is ordinarily conceived of by company law as being an association of people who form a contract between them when creating a company. Therefore, the company's constitution constitutes a contract between all of the members of that company. The company acts through its directors or through people who have powers delegated to them by its directors. The company may only carry out the powers set out in its constitution: after 2006, the company's constitution is contained in a document known as the *articles of association* (s 17 CA 2006). Neither the company nor its directors may do anything beyond those powers. Similarly, the internal rules of the company – how meetings are conducted, rules as to the conduct of business internally – are also prescribed by the articles of association. A standard set of articles created further to the Companies Act 2006 (known as the Model Articles) stand for a company's articles of association unless the company actually has a specific article dealing with any particular issue. In essence, the Model Articles are a default setting for situations in which a company may have inadequate articles of association.

The various legal types of company

Limited liability companies and other companies

Companies may take the form of companies limited by shares, or companies limited by guarantee, or unlimited companies. Unlimited companies offer no limitation of liability to the company's members (s 3(4) CA 2006) and therefore are rarely used. Beyond that, there are two different types of company. A limited liability company exists if its constitution provides that it is a 'limited company' (s 3(1) CA 2006). That a company is a limited company which is 'limited by shares' means that its shareholders have their liability for any losses of the company limited to the amount specified in the company's constitution: typically, this is limited to an amount of £1 per share or something of that sort. Alternatively, a company may be 'limited by guarantee', which means that each member of the company agrees to contribute by way of a guarantee an amount of money identified in the company's constitution to meet the company's obligations. The liabilities of those members are limited to the size of the guarantee. Companies limited by guarantee do not have shareholders nor do they have shares (s 5 CA 2006); instead the members are answerable for the amounts of their guarantees only but do not have ordinary shareholders' rights precisely because they do not own shares. It is common to use companies limited by guarantee to create charitable companies.

Our principal focus in this book will be on the company limited by shares; that is, the company that is most familiar to us all, in that it has shareholders with rights to share in the profits of the company. *Unless specified to the contrary, you should assume that we are always discussing companies limited by shares.* The term 'members' is used frequently in the companies legislation to refer to any person who is a member of a company limited by shares or a company limited by guarantee or any other form of company. In relation to companies limited by shares we can think of the term 'member' as being synonymous with the term 'shareholder', although in those types of company in which there are no shares there can only be 'members' and not 'shareholders'.

Public and private companies

Companies limited by shares may be *private companies* (identifiable by the suffix 'ltd' after their name) or they may be *public companies* (identifiable by the suffix 'plc' after their name). The key differences are that public companies may have their shares and other securities offered for sale to the public, whereas private companies may not have their securities offered for sale to the public (s 755 CA 2006); public companies must also have a minimum amount of share capital (s 761 CA 2006). Throughout the Companies Act 2006 there are different rules dealing with private companies and public companies, most particularly the two provisions just mentioned from Part 20 of the Act. Consequently, the shares that are quoted on the London Stock Exchange and other regulated markets are by definition the shares of public companies. There are many specific rules dealing with the shares of public companies, particularly rules for the protection of the general investing public, which are considered in Chapter 11. This means that private companies cannot have access to those general capital markets. It is provided in s 4(1) CA 2006 that a private company is any company which is not a public company. Public companies must be registered as public companies and must be certified as being public companies (s 4(2) CA 2006).

Community interest companies

There is another form of company – the 'community interest company' – that is formed for social enterprise (s 6 CA 2006). It was created by the CA 2006. This is somewhat odd given that there have been mechanisms for social enterprise in existence for centuries in this jurisdiction, as in industrial and provident societies and co-operatives (discussed in Hudson 2009b, Chapters 51–54). This indicates another feature of corporate law: that governments will attempt periodically to reinvent the wheel so that they appear to be doing something proactive in our social and in our commercial life.

The creation of a company

The number of people required to form a company

A company may be formed by one person or by a number of people (s 7(1) CA 2006). This is a significant development from the *Salomon* principle, which, under the now-repealed 1862 Companies Act, required that there had to be at least seven people forming the company. The modern version of the company recognises that a single individual may wish to create a company as a cypher for herself so that her personal liabilities for the business's activities are limited. Pursued to its end point, this is the logical conclusion to the corporate personality principle in *Salomon*. It was a new development in the Companies Act 2006 that public companies could be created by a single individual. There is a problem in that the public might acquire shares in a company operated by one individual on the basis that many inexperienced investors may assume that a public company is a large organisation as opposed to being a single individual's plaything. Of course, companies can also be formed by infinite numbers of people, but the development in 2006 was in reducing the number necessary to create a public company. However, a company may not be formed for an 'unlawful purpose' (s 7(2) CA 2006).

ıe mechanics for the formation of a company

The mechanics for the formation of a company are as follows. The people who are creating the company must subscribe their names to a memorandum of association and then comply with the registration requirements in sections 9 through 13 (s 7(1) CA 2006). The initial subscribers to the memorandum are the founding members of the company. In a company limited by shares, those subscribers acquire their shares in consideration for paying the amount specified in the company's constitution. The memorandum of association used to be a very significant part of the company's constitution in that it set out the company's objects: that means, it specified the extent of the company's powers, whereas the articles of association principally set out the company's internal rules but little more. After 2006, the articles of association became the company's constitution, except that a memorandum of association is still required further to s 8 CA 2006 as a statement that the subscribers wish to form a company and to become members of that company. However, a memorandum is no longer required to contain the company's objects.

The registration requirements are set out as follows. The memorandum of association in prescribed form must be delivered to the registrar of companies, together with the application form and a statement of compliance (s 9 CA 2006). In future it is expected that it will be possible to carry out this application process on line. The statement of compliance is a statement that the documentation delivered to the registrar complies with the Companies Act (s 13 CA 2006). Section 9 CA 2006 requires a range of documentation to be delivered to the registrar. This further documentation includes a statement of the company's capital and initial shareholdings, which sets out the numbers of shares, their nominal value, and so forth (s 10 CA 2006). A statement of the company's proposed officers is also required, including a statement of the identity of the directors and the company secretary (both of which are explained below) (s 12 CA 2006).

If the registrar is satisfied that the documentation complies with the Act, the registrar will register the undertaking as a company (s 14 CA 2006). The effect of registration is that the subscribers collectively (together with anyone who later becomes a member) become a body corporate: that is, their distinct personalities for this purpose merge into a single legal person in the form of the new company (s 16 CA 2006). Then the registrar issues a certificate of incorporation (s 15 CA 2006).

The company's constitution

The nature of the company's constitution in the articles of association

A company's constitution is contained in its articles of association and any resolutions of the company that may amend those articles from time to time (s 17 CA 2006). A lot of rot is talked about company constitutions. They are perfectly ordinary facets of English law if you understand where they have come from historically. If you are an undergraduate reading this, then you will probably have studied property law as part of a qualifying law degree, as we discussed in Chapter 1, and therefore you will undoubtedly have studied the law on unincorporated associations and how they raise problems of the beneficiary principle in express trusts law. An example of an unincorporated association is a club or society, such as a student law society at university that students join on payment of a subscription, which means that they all agree to be bound by the club's constitution. The constitution forms a contract between the club's members. It is a perfectly ordinary contract in that the members of the association pay money as consideration for the benefits of their mutual association. The modern company grew out of this structure, as discussed in Chapter 1.

When a company is formed, the subscribing members are forming an association between themselves, albeit one with legal personality. The company's constitution provides the terms of that contract, and instead of paying subscriptions (as in a club), the members acquire shares on payment of a consideration specified in the constitution. There is nothing unusual about this structure. Modern company law is still based on the idea that the members form a contract between themselves. Consequently, it is provided by s 33(1) CA 2006 that:

> The provisions of a company's constitution bind the company and its members to the same extent as if there were covenants on the part of the company and of each member to observe those provisions.

Therefore, the company itself as a distinct legal person, and also all of the members of the company are bound by the constitution as if it created 'covenants' (or contractual obligations) between them all. The constitution therefore binds each member to each other member and, significantly, binds the members to their obligations to the company; and vice versa the company is bound to the members. Therefore, it would be possible for two women to create a company on the basis that both of them as its only shareholders would each pay £1 million in equity capital and each provide identified machinery for the business by putting those requirements into the articles of association, such that the company would not move that machinery out of the country: the shareholders would therefore be bound to pay that money and to provide that machinery, and the company itself (perhaps acting through one of the shareholders) would be prevented from moving the machinery to France. Just as the contract in the example of the student law society dictates under contract law what the officers and members of that unincorporated association can and cannot do (*Re Bucks Constabulary Fund No 2* (1979)), so the articles of association bind the company (and therefore its directors indirectly) and the shareholders through this adaptation of contract law principles to fit companies. Furthermore, any amount owed by a shareholder to the company under the articles of association (for example, in payment for shares) is treated as an ordinary debt owed by that shareholder to the company (s 33(2) CA 2006).

The members are entitled to receive a full copy of the current constitution of the company by requesting it from the company (s 32 CA 2006).

The memorandum and articles of association

Before 2006, the company's constitution comprised a memorandum of association and articles of association. The memorandum of association set out the company's objects, and consequently constituted a statement of the limits on the company's capacity to act, whereas the articles of association contained the rules by which the company operated internally. As a result of the Companies Act 2006, the memorandum of association is simply part of the application process when forming a new company, whereas the articles of association now stand for the whole of the company's constitution. All companies are required to have articles of association (s 18(1) CA 2006). The articles of association may be drafted in specific form for each particular company (provided that they comply with any mandatory rules of company law); or alternatively, if there are no articles or inadequate articles, the Model Articles created by the Secretary of State further to s 19 CA 2006 will stand in for those articles (if there are none), or for any gaps (if there are articles but with gaps) further to s 20(1) CA 2006. The role of the articles, then, is to provide for regulations as to the operation of the company (s 18 CA 2006).

For companies created before the 2006 Act came into effect, the old Table A (as it was known) stood for the articles of association as default articles. Now we have the Model Articles instead. Such companies created before the Act came into force used to have the memorandum of association containing much of their constitution, although the effect of the 2006 Act is to treat such provisions of the memorandum as being now contained in the articles of association, so that all provisions in the Act relating to articles of association apply to those old memoranda of association instead (s 28 CA 2006).

Amending the articles of association

The articles of association may be amended by a special resolution of the company (s 21 CA 2006). That means the articles can only be changed if shareholders holding at least 75 per cent of the voting rights in the company approve the amendment. (The rules governing meetings and resolutions are considered in Chapter 6.) However, there may be particular provisions of the articles that the company's members wish to have 'entrenched' from the formation of the company so that they cannot be changed simply by a special resolution of the company. Such entrenched articles can only be amended if the conditions identified in the articles for their amendment are satisfied (s 22 CA 2006). So, for example, in relation to a family company it may be the wish of the family members that the company maintains and operates land as an organic farm in accordance with many years of family tradition. As such, the articles of association may entrench a requirement that the land be operated only as a farm and not for any other purpose unless the farm business is insolvent and all of the shareholders personally are in receipt of bankruptcy petitions: in this way, the company's activities can be entrenched except for circumstances in which identified conditions are satisfied. The existence of such entrenched articles must be notified to the registrar (s 23 CA 2006). Furthermore, the articles cannot be amended so as to compel an existing member of the company to subscribe for more shares in the company (s 25 CA 2006). In all circumstances, any amendments to the company's constitution must be notified to the registrar (ss 34, 35 CA 2006, and so on).

Corporate capacity and authority to bind the company

Companies operate on the basis of their constitutions. To return to our metaphor (in the Introduction to this book) of a company being like Frankenstein's monster, that creation can do anything that its creator enables it to do, but it cannot do anything that its creator does not empower it to do. It is the same with companies. Companies can only do the things that their constitutions permit them to do. In legal parlance, the company only has the *capacity* to do the things that it is permitted to do by its articles of association. The company's members may deliberately limit the company's capacity by placing limitations on it in the articles of association; alternatively they may decide to give the company's directors freedom by placing no restrictions on its activities in the articles of association. The term 'capacity' of a company means its legal powers to do any particular thing. A company without the power to enter into a transaction does not have the capacity to enter into it.

There is a separate, but related, question as to the *authority* of a company's agents to bind the company. Companies act through the human beings who people them. Therefore, the directors and other officers of the company will be delegated with the authority to do specific acts on behalf of the company (to create contracts, to buy land, and so forth). These human beings are therefore the agents of the company. The authority to bind the company will ultimately be governed by the terms of the company's constitution. Therefore, the company's directors, employees and other agents will have their ability to bind the company limited by the terms of their authority, which will flow from the articles of association and any further, particular limitations that are placed on them individually. This is a separate question from the capacity of the company to act, because the company may potentially be able to do something, but a particular agent may be precluded from binding the company to that act even if the company could conceivably have performed it (perhaps because that agent is considered too junior, or the risks too large for one person to decide alone, or whatever).

The problem that these separate questions of capacity and authority create is typically a question as to the rights of outsiders (that is, third parties who are not the company and not its agents) to enforce their rights against the company if the company was acting beyond its capacity or if its agents were acting beyond their authority. These questions are considered in the paragraphs to follow. (None of these provisions applies to charitable companies (s 42 CA 2006).)

The company's objects and the ultra vires principle

A company's 'objects' (or powers) are said by s 31 CA 2006 to be 'unrestricted', which means that a company may do anything lawful unless the articles of association expressly prohibit the company from doing that thing. Previously the law had assumed that companies could only do the things that were expressly permitted by its constitution: therefore, the 2006 regime is more permissive in this regard by assuming that a company can do anything. As a result, in the ordinary course of events a company may borrow money, conduct any business it wishes, buy and sell land, and so on and so on, without any restriction. It is only if the people creating the company (or the members latterly by an amendment of the articles) wanted to ensure that the company only pursued limited activities or avoided risky activities, that the company's constitution would need explicitly to prohibit such activities. The importance of this provision can only be understood if we think about the position before this principle was introduced.

Before the CA 2006 came into force, the objects or purposes of the company were set out in the company's memorandum of association. There was a large amount of case law on the effect of the company's 'objects clause' in the memorandum on the legal effect of its activities. The most pressing problem in practice was that a company created with an object of conducting farming might in time become a company which in fact ran a hotel because the farm ran into difficulties and the picturesque farmhouse struck the directors as being a better business prospect as a hotel. If the hotel business ran into financial difficulties and the directors wanted to escape from contracts which had committed them (for example) to buying new beds for all of the rooms, the directors might have sought to argue that the activity of running a hotel was beyond the company's powers because it was not sanctioned by the company's objects clause in the memorandum of association. Therefore, acting as a hotel would be said to be beyond the powers of the company (or, ultra vires) and therefore a void act. Consequently, it would be argued that the contract to buy the beds was void (such that the company would not have to pay for the beds) because it was ultra vires. This had the effect that third parties would be unable to enforce contracts against the company in circumstances in which they would typically have no idea that the company's objects clause was limited in that way. Nevertheless, the old case law took the view in general terms that those third parties had constructive notice of the contents of the memorandum of association filed with the registrar and therefore that they were bound by the objects clause (*Ashbury Railway Carriage & Iron Co v Hector Riche* (1874–75)).

Therefore, CA 2006 contains s 31 to provide that a company has *unrestricted* objects in the ordinary course of events. To prevent the ultra vires argument being used by a company, there is a provision in s 39 (a principle which was first introduced in 1989) to the effect that the company cannot rely on any limitations in its own objects clause. Section 39(1) provides that:

> The validity of an act done by a company shall not be called into question on the ground of lack of capacity by reason of anything in the company's constitution.

Thus, s 39(1) seeks to protect outsiders against provisions in the company's constitution restricting its capacity being used to call any act of the company 'into question'. The idea of 'calling an act into question' includes arguing that that act is unenforceable or void. Therefore, in relation to *third parties dealing with the company*, there is reassurance that nothing in the company's articles of association (including its memorandum) may be relied upon by the company to set aside a transaction or any other thing on the grounds that it was not permitted by the company's objects. As a result, even if the members of the company had deliberately put something in the articles of association to limit the actions of its directors, that will not be effective against outsiders dealing with the company.

There are some outstanding issues, however. First, the transaction may be set aside on other legal grounds (undue influence, duress or whatever) but not on the ground that it was outside the company's powers. So, the company is still entitled to set aside a transaction on other legal grounds (s 39(3) CA 2006). Second, the agents who have created this transaction do not escape their own liabilities to compensate the company under the general law (s 40(5) CA 2006). However, that will not necessarily recoup all of the company's loss if the agent has insufficient personal funds to meet the loss.

Third, there is the separate question as to whether or not the directors are able to bind the company if they are doing something which the articles of association do not permit: after all, that is an example of an agent acting ultra vires – outside the powers delegated to her – and so the company might argue that the third party ought not to be able to bind the company. This question is considered next.

The ultra vires principle and the authority of the directors to bind the company

Following on from the preceding discussion, it might be said that the directors or other employees or agents who purport to bind the company into a transaction that is beyond its constitutional powers have simply acted beyond their authority and so have failed to create a valid contract. Alternatively, the issue may be that those agents of the company have acted beyond their own personal authority to bind the company. An act may be beyond that person's authority because of a provision in the articles, or because of a resolution of the company, or because of an agreement between the company's members (s 40(3) CA 2006). The question then arises: can third parties enforce a transaction against the company that is beyond the authority of the agent who purported to make it? Section 40(1) CA 2006 deals with this eventuality:

> In favour of a person dealing with a company in good faith, the power of the directors to bind the company, or authorise others to do so, is deemed to be free of any limitation under the company's constitution.

Consequently, there are four requirements for a third party to be able to enforce the transaction. First, the third party must be dealing with the company in good faith. Importantly, it is provided that a person is not deemed to be acting in bad faith simply because she has knowledge that the act is ultra vires (s 40(2)(b)(iii)). (This is an odd provision, given that in equity generally a person is deemed to be acting unconscionably if she has constructive knowledge of something, let alone actual knowledge of it.) Significantly, third parties are not required nor expected to inquire into the extent of the agent's powers to bind the company (s 40(2)(b)(i)). Therefore, one cannot be blamed for failing to find out whether or not an agent had the authority to bind the company. Furthermore, third parties are presumed to be acting in good faith unless the contrary can be proved – so there is no positive obligation on the third party to enquire into the agent's bona fides (s 40(2)(b)(ii)).

Second, the assumption is made for the purpose of this provision that the company's constitution does not contain the limitation on the company's powers. Third, the directors can delegate power to another person to act on behalf of the company, and the principle will still apply. It is suggested that even if this delegation is beyond the powers of the directors (for example, if the transaction is one which the articles permit only directors to create), then that ultra vires act of delegation must also be overlooked under s 40. Fourth, impliedly, the transaction must be enforceable in law generally, beyond the ultra vires principle discussed here (for example, there must be a valid contract in any event). Importantly, the members of the company are permitted to bring proceedings in an attempt to stop the unauthorised act from being performed, as discussed in Chapter 8 (s 40(4) CA 2006).

The question arises as to what is meant in s 40(1) by 'the directors'. Does that mean all of the directors forming a quorum and voting at a meeting of the board of directors, or is it enough that there is an action by only one director or by a small, non-quorate group of directors? Notably, it has been held that the third party need not be relying on an action of the entire board of directors, but rather the actions of individual directors are sufficient (*Smith v Henniker-Major* (2002)). This approach is reflected in s 40 of the 2006 Act. Older authority had required that the directors involved could be demonstrated to have ostensible authority to bind the company: that is, the courts required that it would reasonably appear in the circumstances that the directors were empowered to do what they purported to do, and not that they were acting in such a way that it could not reasonably be thought that they had the authority to do what they were doing, for example if they purported to sell off an office block for only £1 (*Freeman & Lockyer v Buckhurst Park* (1964)). Nevertheless, it has been held that s 40 is not complied with by a meeting of the directors when not all directors were present because the proper procedure for calling the meeting had not been carried out, such that it could not be said that all the directors had agreed as opposed to merely some of the directors insufficient to form a quorum (*Ford v Polymer Vision Ltd* (2009)).

The further question arises as to what is meant by a person 'dealing with' the company under s 40. It is provided in s 40(2)(a) CA 2006 that 'a person "deals with" a company if he is a party to any transaction or other act to which the company is a party'. Simply receiving shares from a company does not constitute a dealing with the company in this sense (*EIC Services v Phipps* (2004)). Otherwise, any act towards a transaction or contract would seem to constitute dealing. What is more difficult to know is whether simply writing letters or attending meetings with the company's officers constitutes dealing, or whether it must amount to something in the formation of a contract or the suffering of detriment by the claimant.

Naturally, s 40 is hemmed in by exceptions in s 41 CA 2006 because it is susceptible to a simple form of abuse. Suppose a director wanted to bleed money out of the company. She might do this by setting up a new company and purporting to create a contract outside the terms of her authority between that new company and the old company such that the old company is required to pay the new company a very high price for otherwise worthless chattels. The director would argue that the new company is entitled to enforce the transaction under s 40 in spite of her breach of her authority. (The director would be liable to hold her profits on constructive trust as discussed in Chapter 7 (*Regal v Gulliver* (1942).) Section 41 deals with these sorts of schemes and provides as follows:

(1) This section applies to a transaction if or to the extent that its validity depends on section 40 (power of directors deemed to be free of limitations under company's constitution in favour of person dealing with company in good faith). . . .

(2) Where –
   (a) a company enters into such a transaction, and
   (b) the parties to the transaction include –
      (i) a director of the company or of its holding company, or
      (ii) a person connected with any such director,
      the transaction is voidable at the instance of the company.

Therefore, a transaction will not be validated solely by s 40 if it falls within s 41. In essence, if a director of that company (or a company that owns the company in question) deals with a person with whom she has a 'connection', the transaction is capable of being declared void. A person is connected to a director if their relationship falls within ss 252 or 253 CA 2006, a list that includes family members and companies owned or controlled by the director. This provision means that the transaction is not automatically void, so that the company may decide, for example, to affirm the transaction if it wishes to do so. Thus, s 41(4) CA 2006 provides that:

(4) The transaction ceases to be voidable if –
   (a) restitution of any money or other asset which was the subject matter of the transaction is no longer possible, or
   (b) the company is indemnified for any loss or damage resulting from the transaction, or
   (c) rights acquired bona fide for value and without actual notice of the directors' exceeding their powers by a person who is not party to the transaction would be affected by the avoidance, or
   (d) the transaction is affirmed by the company.

There are, therefore, practical legal limits on the avoidance of the transaction.

Corporate responsibility

Introduction

This section considers the important question of the circumstances in which the company will bear responsibility for the acts of its employees or its agents for the commission of wrongs, whether in tort or under the criminal law. This topic illustrates some further cracks in the *Salomon* principle that the company is a person distinct from the human beings who work for it, but the question more precisely in this context is whether the acts of such a human being should be treated as being the acts solely of that human being or instead as being the acts of the company itself. After all, if a company can only act through the agency of human beings, then any act of the company itself as a legal person will require a human being to carry out that act. In essence, the question can be thought of the other way around: when can a company be blamed for something that a human being purports to do in its name? We will consider the general principle of attribution of knowledge and fault, before considering the tortious and criminal law liability of companies.

The directing mind and will – attribution of knowledge to the company

Even though the company is a distinct person and even though in this chapter we will consider how decisions that are reached properly will become 'acts of the company', nevertheless there are situations in which the knowledge and behaviour of the human beings who operate the company will be attributed to the company itself. In this way, company law has a concept of the 'directing mind and will' of the company. Its purpose is to plug a very simple gap in the logic of company law: given that so much of our substantive law depends upon the knowledge or dishonesty or reasonableness of a defendant, it is impossible to prove that an abstract entity like a company could have had any of those states of mind. Therefore, the company is generally fixed with the state of mind of which-ever person or people are the directing mind and will of the company. Consequently, a company is said to know anything that a human being who was part of its directing mind and will knew, and so forth. Deciding who is the directing mind and will requires the court to consider who was operating in that capacity from transaction to transaction.

So, where in *El Ajou v Dollar Land Holdings* (1994) the non-executive chairman of a company organised that thieves would invest stolen money in the company, it was held by the Court of Appeal that the chairman had been the directing mind and will of the company in relation to that specific transaction. This was so even though the chairman was not ordinarily in control of the day-to-day management of all of the company's affairs. Consequently, the chairman's knowledge of the theft would be attributed to the company because the chairman was the company's directing mind for the purposes of that deal.

In many circumstances, the directing mind of the company will be one or more of its directors. However, it is possible that some companies will in fact be run by clandestine figures who do not sit on the board of directors but who nevertheless have some influence over the people who do. The position of these 'shadow directors' is discussed below.

Company law, therefore, having assembled the fiction of the company in the first place, has assembled a second fiction (attributing the mental state of a human being to a corporate entity) so as to close that logical gap. In a sense this principle breaks the *Salomon* principle because it looks behind the corporate personality and fixes liability via individual human beings in various situations instead. As suggested in Chapter 2, there are clearly a number of shortcomings with the *Salomon* principle that are assuaged by this doctrine – as is discussed in Chapter 7.

Agency and the identification of the directing mind in complex companies

The attribution of responsibility in large companies

The mainstream principle therefore relates to the knowledge that can be attributed from a person to a company. There is another question: in which circumstances will a human being's actions, nominally on behalf of the company, be considered to be so outré that they will not bind the company? As Lord Reid put it in *Tesco Supermarkets v Nattrass* (1972):

A living person has a mind which can have knowledge or intention or be negligent and he had hands to carry out his intentions. A corporation has none of these: it must act through living persons, though not always one and the same person. Then the person who acts is not speaking or acting for the company. He is acting as the company. . . . If he had a guilty mind then that guilt is the guilt of the company. It must be a question of law whether, once the facts have been ascertained, a person in doing particular things is to be regarded as the company or merely as the company's agent.

There is, therefore, a tension here: in some circumstances, the human being may be acting *as* the company itself, or in other circumstances the human being may be acting merely as the agent of the company. So, if the single directing mind and will of the company commits a fraud, then that will be treated as a fraud committed by the company (*Mahmud BCCI SA* (1998)).

In *Tesco Supermarkets v Nattrass* (1972) the manager of a particular branch of a Tesco supermarket had caused an offence to be committed by allowing goods to be advertised as being offered on a discount. Even though the store had run out of the cheap stock and so had begun to replace it with goods at the higher price, it had not taken down the 'discount' signs, which was an offence under the trades descriptions legislation. The question was whether or not the fault of that branch manager could be attributed to the company. It was held that a branch manager in such a large company, with many stores, could not be part of the directing mind and will of the company because a branch manager would be too junior within the scope of the organisation as a whole. Therefore, the offence could not be attributed to the company.

This does raise the problem that it will be more difficult to attribute liability to large companies simply due to their size because even middle managers are likely to appear to be too junior in the context of the entire organisation to attribute liability to the entire organisation. Consequently, particular regulatory contexts, such as the Financial Services Authority's regulation of financial institutions through the SYS rulebook within the FSA Handbook, impose obligations on companies to construct internal control systems precisely so that they will be responsible for the actions of any of their employees or agents. Otherwise, the common law rules on vicarious liability in tort or the law of agency will have to serve to attribute fault to companies. The question that arises in that situation is the following: what if that employee or agent is exceeding her authority and doing things that the company would never have condoned? Can the company be blamed for those actions?

Agency and companies used as cyphers

In *Stone Rolls v Moore Stephens* (2009) (which was discussed in Chapter 2), the question arose whether or not the fraudulent actions of the directing mind of a company could be attributed to the company because, it was argued, the company could not have authorised any of its agents to have acted fraudulently. Based on the decision in *Hampshire Land* (1896), it was argued that knowledge should not be attributed to a company where it would be contrary to 'common sense and justice' to do so because the agent had, for example, sought to conceal those matters from the company. On the facts of *Stone Rolls* it was held by the majority of the House of Lords that the directing mind and will of the company had, in effect, used the company as a cypher for himself in committing fraud, and therefore no distinction would be drawn between that individual and the company. In that case, an individual committed a fraud by using the company that he controlled to borrow money from banks, and then the company sought to recover damages from its auditors for failing to spot that fraud. The company was prevented from recovering damages from its auditors for failing to discover the fraud of its human operator on the basis that the company and that individual were treated as being one and the same person for this purpose. Lord Phillips considered that the facts of that case made the problem an easy one to solve because that individual was really using the company as a cypher for his own activities.

By contrast, a company with a large number of shareholders and a large number of genuine directors with a substantial business might have been more difficult to blame for the activities of one individual, unless that individual could be shown to have been the directing mind and will of the company in relation to a sufficiently significant transaction. Clearly this area of law intrudes significantly on the *Salomon* principle by permitting the attribution of responsibility from the actions of individuals to the company. This will be desirable in relation to claimants seeking high damages where the company is the only defendant likely to have sufficient money to compensate their loss, or where the activities of a company need to be blamed to prevent its unlawful practices. So in the following sections we consider corporate liability in tort and in criminal law in outline.

The liability of companies in tort

One important context in which companies may be held to be liable is in relation to torts committed by the human beings who operate the company. The leading case of *Lennard's Carrying Co. v Asiatic Petroleum* (1915) set out the core principle in relation to the liability of companies for the wrongs of their human operators:

> [A] corporation is an abstraction. It has no mind of its own any more than it has a body of its own; its active and directing will must consequently be sought in the person of somebody who for some purposes may be called an agent, but is really the directing mind and will of the corporation, the very ego and centre of the personality of corporation.

Which person or people constitute the directing mind and will is something that will depend on the circumstances. Sometimes the will of the company is expressed by the shareholders voting in a general meeting, or by the directors acting as a board, or by a clearly dominant executive taking decisions on behalf of the entire company. In *Lennard's Carrying Co.*, there was just such a single, dominant executive in relation to the registration of an unseaworthy ship, and his actions were therefore deemed by the court to be the actions of the company, such that the company could be held liable for any wrongs that flowed from that action. This idea is important. Rather than being deemed to be the act of an agent for which the company is held to be vicariously liable, what is being said is that if the directing mind of the company performs an act, then that should be deemed to be something that the company did itself.

In the important Privy Council decision in *Meridian Global Funds v Securities Commission* (1995) it considered an attempted takeover of a cash-rich company, ENG, by two individuals. Those individuals were senior executives of Meridian but they acted without the knowledge of the board of directors. The individuals intended that they would pay for the takeover by using Meridian's money in the short term and then use ENG's own money to repay that money after the takeover had been completed. Once the shares were acquired, Meridian became a substantial shareholder in ENG. The two individuals failed to give the notice that was required by the applicable securities regulations once they had acquired 5 per cent of the shares in ENG. The question that arose was whether the knowledge held by those individuals that they should have given such a notice could be attributed to Meridian.

Lord Hoffmann identified four sources of knowledge which could be attributed to companies: where the articles of association supplied it; where specific provisions of company law required it; by means of the principles of agency law such that the knowledge of an agent may be attributed to its principal; and by means of principles of vicarious liability at common law. His Lordship suggested caution in using the 'directing mind and will' test too mechanically: rather, each transaction should be considered on its own circumstances. On the facts of that specific case it was held that the company 'knows it has become a substantial security holder when that is known to the person who had authority to do the deal'. So, when the appropriate company officer knew that enough shares had been acquired to require a notice to be made, then the company would be deemed to know it. On those facts, the assertion that the individuals had conducted this transaction covertly (such that the company might not have been able to find out about it) should not prevent knowledge being attributed to the company.

The liability of companies in criminal law

There are a number of contexts in which companies can commit crimes, typically because legislation creating a criminal offence identifies the means by which it can be attributed to companies. The most important single body of law containing criminal offences for companies is health and safety at work legislation, which provides for a number of different types of offence if the workplace does not meet specified levels of hygiene, safety and so forth. The detail of those offences is beyond the scope of this book. Another significant body of law creating offences, as discussed variously in this book, is company law and finance law, where each require certain filings and registrations of different sorts of information, any breach of which imposes liability on the corporate entity involved as well as its directors. In older cases such as *DPP v Kent & Sussex Contractors Ltd* (1944), the company's employees used false documents to acquire petrol during post-war rationing, which was a criminal offence: the court held that for this purpose those employees were acting as the company (not simply on behalf of the company) and therefore the company was liable for the criminal penalties. Common sense was applied in *R v ICR Haulage* (1944) to identify that there would be some crimes that could not be pinned on a company, such as bigamy, because companies cannot marry.

The most contentious aspect of criminal law relating to companies was always whether or not companies could be considered to be liable for deaths, whether on account of murder or manslaughter. The principal argument against such liability was that a company could not sensibly form an intention to commit murder, nor any of the various aspects of *mens rea* required for manslaughter. Instead, it was argued that it was better to focus on the human beings who had caused such a death. However, given the difficulty of attributing fault in complex corporate contexts where various people may have played a small part in creating procedures or practices that led to a death, it was argued by others that liability should fall on companies in some circumstances. Indeed, it was argued that it was a moral imperative that corporate entities, and by extension their management, staff and shareholders, should face responsibility for their failings. More prosaically, it also means that the company's assets can be attacked by a fine.

The Corporate Manslaughter and Corporate Homicide Act 2007 embodied the principle that a company may be liable for manslaughter or homicide. The offence is set out in s 1(1) of that Act in the following terms:

(1) An organisation to which this section applies is guilty of an offence if the way in which its activities are managed or organised –
    (a) causes a person's death, and
    (b) amounts to a gross breach of a relevant duty of care owed by the organisation to the deceased.

The organisations that are covered by this provision include corporations (such as companies), a police force and a partnership that acts as an employer. There is an important proviso to this offence in s 1(3):

(3) An organisation is guilty of an offence under this section only if the way in which its activities are managed or organised by its senior management is a substantial element in the breach referred to in subsection (1).

Therefore, the three key elements of the offence are as follows. First, it is the organisation that is guilty of the offence. Therefore, the directors and the shareholders are not guilty of homicide by virtue of this offence, only the company.

Second, the offence is committed if the management or organisation of the organisation's affairs is either the cause of death or constitutes a 'gross' breach of a duty of care owed to the deceased. Therefore, if the death was caused by a freak gust of wind blowing a customer off a company's fairground ride, that would not impose liability on the company, unless there was a defective safety bar on the ride that allowed the customer to be blown from the ride. The 'relevant duty of care' is defined in s 2 to encompass duties owed to employees, duties as an occupier of premises, duties arising in various ways through its ordinary trading activities, and duties owed to anyone whose safety depends on the company.

Third, it must be a 'substantial element' of that breach that 'senior management' managed or organised 'its activities' in such a way that it led to the offence. Therefore, if the death was caused by the negligence of an employee due to him being tired that day, that would not have resulted from the way the company was run by senior management. However, if the cause of death was a policy by senior management that safety rails should only be repaired once every six months so as to reduce costs, and it was a rail that had been defective for five months that caused a customer to fall to their death, that death would have resulted substantially from a failure of senior management. What is less clear is how tightly drawn the idea of the 'activities' of the company should be: if the activity that caused death was outside the ordinary business of the company, that would seem to avoid liability.

The administration of company business

Corporate acts: contracts, execution of documents

The operation of a company and the valid actions of a company are considered in the sections to follow and in Chapter 6. In this section we are considering the creation of contracts and so forth. A company creates a valid contract either if that contract is created in writing under the company's common seal, or if a person acting with authority to create the contract does create that contract on behalf of the company (s 43 CA 2006). A company executes documents (for example, to transfer land) by affixing its common seal to the document, or by a signature of behalf of the company that is witnessed appropriately (s 44 CA 2006). A common seal is a seal with the name of the company 'engraved in legible characters' on it (s 45(2) CA 2006). A company does not have to have a seal (s 45(1) CA 2006), and instead can act by means of signatures as described above.

The company's name

There are lengthy provisions in Part 5 of the Companies Act 2006 relating to the names that can be used for a company. In essence companies may not use names that would constitute an offence (for example, through being an obscene publication) or that would be offensive (s 53 CA 2006), or that are otherwise sensitive (s 55 CA 2006); and companies may not use company names that already exist (s 66 CA 2006), or that could be confused with company names which already exist (s 67 CA 2006), or that would suggest connections with public bodies (s 54 CA 2006). Objections to company names may be made under s 69 CA 2006. It is an offence to give misleading information in relation to the naming procedure of a company (s 75 CA 2006). A name can be changed by special resolution and by notification to the registrar (s 78 CA 2006).

Public companies are obliged to place the words 'public limited company' or 'plc' after their names (s 58 CA 2006). Private companies are obliged to place the word 'limited' or 'ltd' after their names (s 59 CA 2006). Those provisions also provide for Welsh equivalents ('ccc' and 'cyf' respectively). Some companies, such as charities, are exempt from these provisions (s 60 CA 2006). These requirements are intended signal to third parties that the entity is a company and thus also signal the legal rules that will bind that entity. Consequently, s 82 CA 2006 requires that trading companies are required to display their name with the suffix in s 58 or s 59 CA 2006 as appropriate, and to include it in correspondence and so forth, in accordance with regulations created under s 82(1). Failure to do so without a reasonable excuse is a criminal offence (s 84 CA 2006) and opens up liability to civil claims for any loss so caused (s 83 CA 2006). A company must have a registered office to which all communications and notices may be sent (s 86 CA 2006).

Moving on ...

This chapter has sketched an outline of the structure of companies, its has considered the powers and responsibilities of companies and their constitutions, and it has given a flavour of some of their key personnel. In the next chapter we consider the ownership of a company by the shareholders and the precise meaning of that position, and the nature of a share.

Chapter 5

The ownership of a company

Introduction

The nature of share ownership

The ultimate ownership of a company limited by shares (the most common form of company) can be understood as being divided among its 'shareholders': literally, people who own shares. The way in which company shareholdings operate can be very simple or they can be made wilfully complicated. This chapter sets out a brief description of the various forms of shares, the means by which those shares are issued, and the rights that attach to those shares in the form of dividends. For the time being, we are only interested in understanding how a shareholding operates at a basic level, whereas in Chapter 10 on the company's capital we shall consider more the complexity surrounding the entirety of a company's capital assets and liabilities, and the legal rules that govern them. We shall begin by asking what it means to be a shareholder.

What it means to be a shareholder

Shareholders own companies, albeit in a stylised way. In a commercial sense, they own a company by being the people who are entitled to share in its distributable profits and who are ultimately entitled to the company's property if it is wound up (once the creditors and others have been paid off), as discussed in Chapter 13. In that sense, while the company owns its own property, the shareholders own the company. They do so in proportion to the size of their investment compared to the total share capital in the company. Each shareholder acquires one vote for each voting share in the company at company meetings, and it is through this democratic process that shareholders ultimately exercise power through the company's constitution. However, not all shares are voting shares: some shares may be issued on the basis that the shareholder will be paid a fixed amount of money each year but without a right to vote as a result. Already, we can begin to get the sense that shareholding may subdivide into complex categories. Therefore, what we must do first is to consider carefully what it means to be a shareholder and then consider the different types of share that can be held.

The commercial aspects of share ownership

A share in a public company can be bought and sold freely. Therefore, an investor in a public company may be investing in the hope of a long-term income from that company in the form of annual dividends, or in the hope of a short-term gain from the market value of that share increasing. In relation to private companies, an investor is not able to sell shares on the open market and so will be investing to earn an income, or in the hope of being able to effect a private sale of the shares in accordance with the terms of the company's constitution, or because she has some personal connection with the activities of the company (as with the Gilbert farm example in Chapter 3).

In a public company, the single vote that is carried by each share is often of little practical use to a private shareholder because no single shareholder has sufficient voting power to exercise control as an individual, and so it is a club of so-called 'institutional investors' in relation to public companies that tends to exercise control of voting rights. As a result it is these institutional investors who are consulted closely by management. In the Takeover Code, for example, a shareholding of 30 per cent of the shares in a public company is generally taken to constitute control of such a company.

The shareholder's property rights are expressed solely by ownership of the share as a form of a bundle of rights that can be transferred or turned to account. Shareholding is not a proprietary concept, in that sense, unless and until the company is wound up. That is, a shareholder does not own any of the company's property until the company is wound up and property is allocated to the shareholder in that winding up. Before a winding up, all that the shareholder owns is the share itself. Nor do the shareholders own any of the company's profits until a dividend is declared out of distributable profits. The expectation of a dividend distributed from the company's profits is nothing more than that mere expectation unless and until a dividend is declared by the directors.

The nature of a share in law

Prosaically enough, s 541(1) of the CA 2006 defines a share as being a 'share in the company's share capital'. A share was defined by Farwell J in *Borland's Trustee v Steel Bros. & Co. Ltd* (1901), 288 as being:

> the interest of a shareholder in the company measured by a sum of money, for the purpose of liability in the first place, and of interest in the second, but also consisting of a series of mutual covenants entered into by all the shareholders inter se in accordance with [the Companies Act, s 33]. The contract contained in the articles of association is one of the original incidents of the share.

The articles of association constitute a contract between the shareholders: therefore, one of the important aspects of shareholding is the mix of rights and obligations that are bestowed on the shareholder by the company's constitution. Therefore, beyond the commercial aspects of a share considered above, the share also embodies a web of contractual terms. A share in a company is in itself a form of personal property (CA 2006, s 541).

Different types of shares and the rights attaching to them

There are many different types of shares. 'Ordinary shares' carry one vote per share; they entitle the shareholder to receive a dividend in an equal amount per share as all other shareholders of the same class if a dividend is declared; and they express a right to a share of the company's assets if the company is wound up.

'Preference shares' entitle their 'preferred shareholders' to a fixed dividend. This makes preference shares equivalent to a bond that pays a fixed rate of interest and that carries a subordinate right to participate in the company's property. There is a 'preference' in the sense that the shareholder is guaranteed the receipt of a dividend of a given amount, and is not required to wait in the hope that there will be sufficient profit to afford a dividend on an ordinary share. It is common for a preference share not to carry any voting rights. However, whether preference shares will carry voting rights and whether they will be entitled to a dividend beyond the guaranteed level of dividend will depend upon the terms of the articles of association relating to those shares.

Within large companies it is common to have different 'classes' of shares. This means that, even within the categories of ordinary or preference shares, there will be subcategories (or classes) of such shares. Each class of shares will have particular rights attached to those particular shares. Essentially, a particular class of shares will have particular rights attached to it because that was necessary to induce the target investors (whether small investors or institutional investors) to invest in the market conditions at the time that those shares were issued.

Between these two principal types of share, there are many possible permutations. They are known generically as 'securities' and are considered next.

Complex securities rights, sometimes involving shares

Corporate finance can be made as complex as you please, but it rests on some basic fundamentals. There are two ways in which a company can acquire capital: it can raise debt or it can issue shares. Debt is acquired in the form of ordinary loans or by means of issuing bonds. In a bond, the investors lend money to the company in return for a payment of interest during the life of the bond and the repayment of the capital sum at the end of the life of the bond. By contrast, raising capital through the issue of ordinary shares does not legally oblige the company to pay anything to the investor. In that sense, the issue of shares presents a lower risk to the company than issuing bonds, although a company which did not pay dividends to its shareholders or which allowed the value of its shares to fall would find it very hard to raise capital through share issues in the future. (See Hudson 2009b, Chapter 33.)

Bonds and shares are both known as 'securities' in market parlance. A transferable security can be traded: that is, the investor can buy or sell securities that have already been issued. The securities of public companies can be traded on public regulated markets and offered to the public, whereas the securities of private companies may not be offered to the public (s 755 CA 2006) and so are more difficult to sell. The issue of any type of security (including shares) by a public company is governed by securities regulations, as discussed in Chapter 11. In market parlance, capital that is raised by means of the issue of a share is known as 'equity capital', although that has no relationship to 'equity' as it is understood by a lawyer.

As an illustration of the potential flexibility in securities issues, there are 'convertible bonds', which begin life as bonds but which can be converted into shares. Again, the precise mechanics of these securities depends upon the terms that are attached to them in the articles of association, including any rights to a dividend and any rights to vote. The power to convert from debt to equity may depend upon the happening of some contingency, or it may be vested entirely in the owner of the security. The type of share that would be acquired on the exercise of the conversion would depend upon the terms of the security issue. For the investor, this offers the benefits of an income from a bond combined with the ability to acquire the rights associated with being a shareholder if that appears to be preferable later on. In recent years a species of security called a 'coco' has developed: that stands for a 'contingent convertible', which means that on the happening of an identified event (such as the financial position of the company worsening in a specified way), the bonds acquired by an investor automatically transfer into shares.

In recent years the growth of financial derivatives has added further dimensions to the sorts of securities that are in issue. Share options are a significant example. An option gives the holder a right to acquire a share (or to sell a share, as appropriate) for a given price on the happening of given conditions. These are popular mechanisms for rewarding employees in companies: they cost nothing for the company in the first place (because no share actually needs to be issued until the option is exercised) and they give the recipient a feeling of being wealthy and connected to the company. Financial derivatives like options also enable financial speculators to gamble on the share price of public companies.

The right to vote

One ordinary share will carry one vote at a company meeting. Company meetings are considered in the next chapter in relation to the operation of a company. However, the right to vote is an important right of a shareholder because changes to the company's constitution, decisions as to the removal or retention of directors, and so on, are all decided upon by votes of shareholders.

The acquisition of shares

Introduction

There are two contexts in which a person may acquire shares: either on the basis of subscription for shares when they are offered for the first time, or by buying shares which have already been issued in the after-market. In this section we will consider the rules on the allotment of shares when they are offered for the first time and vested in subscribers; then we consider the transfer of shares after they have been issued by way of sale or gift.

Allotment of shares

The allotment code in CA 2006

Shares must be allotted to shareholders before they formally acquire rights against the company as shareholders. After all, when shares are offered to the public, it is possible that there will be more subscribers for shares than there are shares to satisfy their orders. Consequently a shareholder cannot become the owner of shares until an identified number of shares are actually allotted to that shareholder. The concept of 'allotment' for these purposes includes the grant of rights to subscribe for ordinary shares or the grant of a right to convert securities into ordinary shares. A new code on the allotment of shares was introduced by Part 17 of the CA 2006 ('the allotment code'). The 2006 allotment code has liberalised the authority of directors to allot shares, provided they are permitted to do so by the articles of association. The heart of the directors' authority to allot shares is encapsulated in the general principle in s 549 CA 2006 to the effect that directors of a company may not issue shares unless the situation falls within one of the broad exemptions from that principle in the statute. A director who knowingly and wilfully contravenes, or permits or authorises a contravention of s 549 of the CA 2006, commits a criminal offence and is therefore liable to a fine under s 549(4) of the CA 2006.

The CA 2006 then provides for a number of exemptions from that provision. There are two particularly significant exemptions, which effectively provide the directors with a power to allot shares. First, s 550 of the CA 2006 provides that the directors of a private company with only one class of share may 'exercise any power of the company' to do any of the three activities related to allotment: namely, allot shares in that company, grant rights to subscribe for shares in that company, or convert any security into a share in that company. However, the directors may not exercise this power if they are 'prohibited from doing so by the company's articles' (s 550 CA 2006).

Second, s 551 of the CA 2006 provides that the directors of a company that (by inference from s 550) is not a private company with only one class of share may 'exercise any power of the company' to do any of the three activities related to allotment: namely, allot shares in that company, grant rights to subscribe for shares in that company, or convert any security into a share in that company. Significantly, however, the directors are only permitted to exercise such a power if they are 'authorised to do so by the company's articles or by resolution of the company' (s 551 CA 2006).

In general terms, the acquisition of shares constitutes an ordinary contract of sale between the company and an original subscriber for shares, or between the seller and buyer of shares that are already in issue. The common liability that may arise in relation to the sale of shares is considered in Chapter 11.

General restrictions on allotment of shares to protect the company's interests

There are restrictions on the allotment of shares that are aimed at protecting both the existing shareholders and the capital base of the company by stopping directors from allotting shares randomly or for ulterior purposes. (The general policy relating to the protection of the capital base of the company is discussed in Chapter 10.)

Directors are prevented from issuing shares to new shareholders or otherwise allotting shares in a way which dilutes the rights of existing shareholders by giving existing shareholders the right to object to a new issue of shares (s 561 CA 2006). This is referred to as a 'right of pre-emption'. Furthermore, those who apply for shares are to be protected to the extent that the capital base of the company will not be increased if the issue is not fully subscribed, or that the company's capital base will only be increased if the terms of the issue envisaged an increase, even if the issue was not fully subscribed (s 578 CA 2006).

The board of directors cannot in any event issue shares beyond the amount fixed as the authorised capital of the company. Furthermore, the directors may issue shares only for a proper purpose: that is, they may only issue shares in the best interests of the company. Directors are required to act in the way that they consider, in good faith, 'would be most likely to promote the success of the company for the benefit of its members as a whole' (s 172(1) CA 2006). An issue for an ulterior motive, such as altering the voting power in the company or for serving only the self-interest of the directors, would be void (e.g. *Hogg v Cramphorn Ltd* (1967), as discussed in Chapter 7).

The transfer of shares

There are two ways of transferring shares: through the traditional certificated form and in the more modern uncertificated form. Section 770(1) of the CA 2006, which continues the policy of ensuring that there is an instrument that can be stamped so as to be subject to stamp duty, provides that it is unlawful for a company to register a transfer of shares unless a 'proper instrument of transfer' has been delivered to the company. If a shareholder has transferred all of the shares comprised in one share certificate to one person, that transfer is effected as follows. First, the transferor sends the transferee the 'proper instrument of transfer' required by s 770(1) of the CA 2006 and executed by the transferor, together with the share certificate relating to the shares comprised in the transfer. Then, the transferee executes the transfer (if it is in accordance with the company's articles), and forwards it to the company to be registered, together with the share certificate and the registration fee.

A transfer may also be registered, at the request of the transferor, in the same manner and subject to the same conditions as if registration were applied for by the transferee (s 772 CA 2006). Ordinarily on a sale of shares the terms of the contract between the parties is that the vendor shall give the purchaser a valid transfer and do all that is required to enable the purchaser to be registered as a member in respect of the shares, the purchaser's duty being to get herself registered (*Skinner v City of London Marine Insce. Corpn.* (1885)).

This traditional means of transferring shares has been superseded by the advent of electronic communication and the Big Bang in London in 1987, when physical trading on the floor of the Stock Exchange was displaced by electronic trading. Consequently today, shares that are listed on the London Stock Exchange are transferred through members of the Exchange, and are settled by means of an electronic transfer through the CREST system, which was introduced in 1996. CREST means that there is no need for either a written instrument of transfer or a share certificate. In that sense, the securities are 'uncertificated'.

Uncertificated securities transfers (whether through CREST or not) are governed by the Uncertificated Securities Regulations 2001 (USR, SI 2001/3755). It is a requirement of the admission of securities to listing on regulated markets and for offers to the public (as discussed in Chapter 11) that those securities are eligible for electronic settlement (Listing Rules, 6.1.23R). Therefore, the shares of all major public companies are in uncertificated form. The regulations 'enable title to units of a security to be evidenced otherwise than by a certificate and transferred otherwise than by a written instrument' (USR, reg. 2). The regulations require that the company's articles of association permit an uncertificated form for that company's securities; or else a special procedure is required if the articles do not permit such a form (USR, reg 15). Once a transfer has taken place, the operator is required to register that change in ownership on its register (USR, reg 27(1)). The issuer and the operator of the system are required to maintain a register of securities (USR, reg 20), and entry on the register is prima facie evidence of ownership of that security (USR, reg 24).

Register of shares

Every company is required to keep a register of its members (s 113 CA 2006). This register of a company with shares must record the names and addresses of shareholders, and it must record the number of shares held by each shareholder.

The register must be available for inspection (s 114 CA 2006). Again, this is a means of providing information to the public about companies.

The shareholder and the company's capital

The important questions about the ways in which companies raise capital from the public (and the securities regulations that govern capital-raising by public companies) are analysed in Chapter 10. The equally important technical questions about the management and maintenance of a company's capital are considered in detail in Chapter 11.

Dividends and distributions

Long-term investors in companies often seek a regular income from their investment, as opposed to short-term investors who hope to earn a profit by selling their shares. An investor who is seeking an income from her shareholding will be seeking 'dividends' from the company, being a payment made to shareholders in an equal amount for all shares of the same class made out of the profits of the company for that accounting period. It is a difficult decision for the directors of a company to decide how much of the profit should be reinvested in the company and how much of the profit should be distributed among the shareholders. A large dividend will encourage shareholders to maintain their investment in the company and will also encourage others to invest in the company in the future. However, a large dividend also depletes the cash that is retained in the business. In some circumstances a group of shareholders in collusion with management may seek to drain a large amount of cash out of the company by means of an exceptional dividend, thus leaving the company with little capital. It is with this last sort of abuse in mind that regulation on the amount of dividend that could be paid out by any company was introduced.

Section 830 CA 2006 requires that any company 'may only make a distribution [to its members] out of profits available for the purpose'. Therefore, the directors must identify the amount of available profit in the relevant accounting period. Accounting practice and the relevant accounting standards identify what will be deemed to be the company's distributable profits in any given situation on a true and fair view of the company's financial position. This limits all distributions from the company to income amounts and prevents dividends being paid out of the company's capital. Furthermore, under s 831 of the CA 2006, after the payment of the dividend the net assets of public companies must still exceed both its share capital and its capital reserves. Thus, the regulations do not specify an amount in numbers that stands for the maximum dividend. Instead, it takes the only possible step of specifying that dividends must be paid out of income, and that they must not deplete the company's capital base. After all, the size of any dividend will be relative to the amount of the company's profits and overall size.

Moving on ...

What we must consider now is the legal context in which companies are managed and operated as we turn to look at the role of the directors within the company and their general duties in Chapter 7. We shall return, however, to consider some of the principal rights of shareholders in Chapter 8.

Chapter 6

The management and operation of a company

Introduction

Understanding company law as creating a legal model with state support

This chapter is concerned with the way in which a company is managed and operated. One way of thinking about the way company law works is as a sort of 'How to . . .' manual. If you want the benefits of limited liability, a structure for your owners to share their rights, a number of means of acquiring further capital and so on, then company law offers you a model for doing all of those things. If you obey the rules governing that model, you will receive the protection of English law. That is effectively how all private law works: the law creates models and you obtain the protection of the law if you follow the rules on creating and operating those models. Company law is one of the most complex examples of private law creating such a model.

So this chapter is a description of a user's manual for a company. It also introduces the most significant section of this book on the duties of directors, the rights of shareholders and the running of companies. We shall consider the following issues: the office of director; the provision of information (including accounts and the audit process); the running of company meetings; and the role of other officials within the company. The provision of information is the principal focus of most company law: ensuring that the company and its operators provide sufficient information of the right kind to third parties, and identifying who is responsible for doing that is central to company law.

The bureaucracy of company law

It is possible that your preconceptions about company law were orientated around commercial deal-making, corporate boardrooms with immaculate wooden tables and professional people with movie-star looks rushing around an open-plan office trying to save the world. In truth, a large part of company law is concerned with bureaucracy: the formalities for establishing companies, the rules that govern shareholder meetings and the preparation of annual report. Hence this chapter is called 'The management and operation of a company' because most of company law is concerned with the detailed, day to day operation of companies.

Much of company law has been created to prevent abuse of the corporate form as well as to create order in its functioning. If you cast your mind back to Chapter 1, a large degree of fraud surrounded companies in their earlier days. Consequently, a great part of company law throughout the 20th century was concerned with ensuring that companies published enough information for the benefit of a wide range of third parties dealing with the company – such as business creditors, banks, other traders, and customers – so that they are able to make informed decisions about the risks associated with that company. In essence, because of the artificiality of the company as an organisation (under which human beings can hide behind limited liability and the fiction that the company itself is a person), it was necessary to impose bureaucratic obligations on the company to publish audited accounts, to publish information about its directors and its shareholders, and latterly to publish information about the condition of the business when dealing on securities markets.

All statutory references are to the Companies Act 2006 unless otherwise stated.

The discussion of 'corporate governance' in this book

So this chapter provides you with an overview of those bureaucratic responsibilities, and an explanation of their significance. In Chapter 7, 'Directors' duties', we shall consider in detail the law governing the general duties of directors that are set out in the statutory code which was introduced by s 170 *et seq.* of the Companies Act 2006 (CA 2006). The counterpart to the duties of directors is the law on the rights of shareholders, which is considered in Chapter 8, 'Shareholder rights'. Then in Chapter 9, 'Corporate governance' we shall consider the way in which non-legal codes of practice have been introduced to govern the way in which large public companies are operated in practice.

Taken together, these chapters constitute a composite analysis of what is known as the 'governance' of companies, achieving a trade-off between the powers and the responsibilities of the directors, the rights of shareholders, and the needs of the company as a distinct legal person. It is important to bear in mind that the logic of company law maintains that the company is itself a distinct person. Therefore, we are not only concerned with the needs of the directors, the needs of the shareholders, or indeed with the needs of any employees, but rather we are also concerned with the needs of the company as a distinct, abstract entity. This is particularly significant if the company is a trading company because the company's business may have long-term investment needs or its position in a market may have a long-term logic to it that is contrary to the short-term wishes of shareholders to take a large dividend or of the directors to increase their pay.

The paradox in company law thinking, and the various types of company

You should bear in mind that there is an important paradox in company law. As far as company law has been concerned, traditionally the directors owe obligations to the company. Most significantly, those obligations are fiduciary in nature (as considered in the next chapter). The shareholders have rights to vote in company meetings and have rights to remove the directors (as discussed below). In essence, then, the directors' fiduciary obligations make the director appear to be the person who bears burdens in relation to the company. However, if one looked at the matrix of powers and obligations in a company from the perspective of a management theorist, then one would see the executive directors of the company as having all of the power to operate the business, to make decisions, to pay a dividend out of distributable profits, and so on. Seen in this light, the director seems to be powerful and the shareholders (and importantly, junior employees) to be subordinate to the directors. Therefore, the company law perspective on the directors' role is paradoxically different from the perspective taken from an employee or a minority shareholder in a company.

The operation and management of public companies

Public companies have access to many forms of funding that are not open to private companies (further to s 755 CA 2006). They therefore have a broader range of management considerations to bear in mind than do private companies in this context, because they must consider the needs of capital markets (for example, the stock markets and bond markets) and the demands of securities regulation. Among such concerns will be the needs of its shareholders, where its shareholders are typically institutional shareholders and a large number of private individuals with small shareholdings. The good name of the company will be important both in relation to the viability of its business and in relation to its ability to raise future financing from capital markets.

Importantly, there are also different types of directors in public companies. In practice, companies will have different grades of director. At the most obvious level, there will usually be full-time executive directors and part-time non-executive directors. In essence, executive directors conduct the day-to-day management of the business while non-executive directors keep an eye on the performance of executive directors in certain contexts. In large public companies good corporate practice generally requires that executive directors may not control all aspects of management (as considered below), especially decisions relating to their own salaries. The movement towards control of directors and management began with concerns about rapidly increasing management pay in large public companies and then took on another dimension with the large corporate failures in Enron, WorldCom and other companies. Those corporate collapses are considered in detail in Chapter 9, 'Corporate governance'.

In essence, in relation to the large public companies that might have the greatest effect on the economy and in which ordinary members of the public (as well as professional investors) may hold shares, the concern was to ensure that those sorts of companies complied with basic corporate governance requirements. Consequently, non-executive directors were introduced to large public companies. Non-executive directors in such companies are generally drawn from backgrounds which give them useful perspectives on the company's business: in fact, we might worry that they are too similar to the executive directors and that effectively a small group of people who share the same belief system are acting as executive and non-executive directors for all of the largest companies in the UK. Non-executive directors are used to decide issues which directors ought not to decide alone (for example, in relation to directors' pay) and are used to advise the executive directors on the most appropriate way for the company to act in a number of circumstances.

Particularly significantly in public companies there is a distinction made usually between the Chief Executive Officer (CEO) and the Chairman, with the powers of each being delineated in the articles of association (as considered below). The CEO is in charge of the day-to-day management of the company. If there has to be one 'boss', then it is the CEO. However, to prevent the CEO (or her cronies) from exercising complete control over the company, with the possibility of causing crashes like that at Enron, the practice developed to have a non-executive Chairman who could exercise control over the CEO. In essence, the Chairman is a counterbalance to the CEO's power. Importantly, it is not company law that requires that this distinction exists, but rather it is the non-binding collection of corporate governance documents described in Chapter 9 that presents it as best practice. To reassure investors, most publicly quoted companies have adopted these structures. Therefore, ordinarily the Chairman will chair committees dealing with the auditing of the company, directors' remuneration and so forth, to reassure outsiders that these issues are properly conducted without being the exclusive preserve of the executive directors and the CEO.

It should not be thought either that all directors are the same or that they all perform the same functions. The more complex the company or group of companies, the more likely it is that different directors will have very different responsibilities within the organisation. Suppose a group of companies produce two different products – widgets and twidgets – and the companies take out loans from banks, they employ manufacturing staff and sales staff in two countries, and their shares are quoted on the Stock Exchange. Clearly, different people will need to be in charge of the widget and twidget business units, in charge of production in the different countries, in charge of marketing, in charge of personnel, and in charge of the financing for the groups of companies. It is probable that there would be at least one different director on the central board of directors to fulfil each of these functions. Depending on the size of the business units, it is more likely that each of these business areas would require a number of directors.

Clearly, controlling and motivating these various actors within a large corporate entity is a demanding task for a CEO. Typically, each business unit would require managing directors who would report to the CEO. The breadth of managers

– whether a large board of directors or a small board of directors – would depend upon the management style of the CEO and the scope of the company's operations. Consequently, there can be differences in the levels of power, influence and remuneration of directors. A number of different types of people may be treated as being a director, as considered next. Company law tends to assume that all boards of directors will act unanimously in all decisions, whereas in truth most groups of companies have their own politics and cliques. Different people will want to be more powerful, better paid, to pursue their vision and to rise to the top in the future. Therefore, the operating practices and articles of association of different companies may be different to accommodate their particular styles. There will always be a distinction between the formalities of company law and commercial practice.

The office of director

Directors and the managers of companies

This section considers who will be a director, the general nature of directorship, and the role of directors in managing a company. The next chapter considers in detail the code of directors' duties set out in s 170 *et seq.* CA 2006. So at this stage we are concerned with identifying who will bear the duties of directors and some of the procedural obligations that are placed on directors by the CA 2006, before Chapter 7 analyses the detail of directors' general duties. Before we come to those important general duties, however, it is important to understand the operation of a company as described in this chapter.

Companies act through their directors

Companies are a fiction. They do not exist. Except, that is, for the fact that human beings treat companies as though they exist, and consequently they have a sort of existence. But because companies have a legal existence, some human beings must be the vehicles through which they act. To make a long story short, those human beings, for most purposes in the companies legislation, are the directors.

The identity of the directors

Who is a director?

Section 250 CA 2006 provides that the term 'director' includes 'any person occupying the position of director, by whatever name called' and therefore it does not matter what label is put on a person by the company itself, because what counts is the function that that person actually performs. So, the straightforward answer to the question 'who is a director?' is that the concept 'director' encompasses anyone who has been expressly appointed as a director in accordance with the articles of association of the company in question. There are, however, other people who may be treated as being directors of a company, and consequently will have the obligations of directorship imposed on them, even though they have not been formally appointed as such. These categories of people are considered next.

Who else will be a director? De facto directors and shadow directors

In most companies the identity of the directors will be self-evident because they will have been appointed as such and will meet as a board of directors on a regular basis. However, there are situations in which another person, or other people, may obtain such an influence over the ordinarily appointed directors that the law will treat them as though they were directors: such a person is known as a 'de facto' director. Another important situation in which it may not be clear whether or not a person is a director is where that person tries to conceal their role within the company: for example, if they have been disqualified from acting as a director or if their participation might have unfortunate tax or regulatory consequences. Such a person is known as a 'shadow director'.

De facto directors

A 'de facto director', broadly put, will be a person who acts as though she is a director, or who holds herself out as being a director, without having been appointed as a director. This may involve a junior employee who is allowed to lead people to believe that she is a director when in fact she is not; or someone who appears to outsiders to be acting as a director without actually being one, for example, because she has been delegated a power to sign contracts on behalf of the company. Clearly, it is a difficult line to draw when someone is or is not acting as a director. In *Re Hydrodam (Corby) Ltd* (1994), 183, Millett J held that:

> A de facto director is a person who assumes to act as a director. He is held out as a director by the company, and claims and purports to be a director, although never actually or validly appointed as such. To establish that a person was a de facto director of a company, it is necessary to plead and prove that he undertook functions in relation to the company that could properly be discharged only by a director. It is not sufficient to show that he was concerned in the management of a company's affairs or undertook tasks in relation to his business which cannot properly be performed by a manager below board level.

Therefore, on this test, one must demonstrate that the individual was doing things that could only be done by a genuine company director (presumably in the context of that particular company). What is important to Millett J is that the director must have been holding herself out as being a director: that is, purporting to be a director.

In *Secretary of State for Trade and Industry v Tjolle* (1998), Jacob J held that a person became a de facto director on the basis both of being held out to be a director, and on the basis that those activities could only be performed by a director or that the defendant was taking major decisions at a directorial level, or was acting on an equal footing with others in directing the affairs of the company. This narrows the test suggested by Millett J. In *Re Kaytech International plc* (1999), the Court of Appeal held that ultimately the court must be satisfied that the individual had assumed the status and functions of a director, so as to have openly exercised real influence in the corporate governance of the company.

In *Holland v HMRC* (2010) Lord Collins held that a de facto director must have assumed the responsibilities of a director and must be part of the company's governing structure. Lord Walker considered that that person must take all of the key decisions in relation to a company and be able to ensure that they are carried out; although this seems relevant only for small companies operated by a few people. (This case is considered at www.alastairhudson.com.)

Shadow directors

Section 251 CA 2006 provides that a shadow director is someone in accordance with whose instructions the directors are accustomed to act. This excludes people who are providing purely professional advice, such as solicitors. A shadow director is therefore someone who has no official status as a director within the company but who tends to give directions to the directors that those directors usually obey. There may be circumstances in which the driving force behind a company wants to remain anonymous. Typically, this would be because the company has been established for regulatory or tax avoidance purposes (as discussed in Chapter 3). Consequently, that person may have genuine influence over the directors but without having any legal power to control them. Such a person would be a 'shadow director'.

Several of the statutory provisions in both the Companies Act 2006 and the Insolvency Act 1986 relating to directors also apply to shadow directors. Shadow directors will also bear fiduciary duties (*Yukong Line Ltd v Rendsburg Investments Corporation of Liberia (No. 2)* (1978), 311). Therefore, a shadow director exercises control in a way that would be invisible to anyone looking at the company's books or structure from the outside. In that sense, they could be thought of as lurking in the shadows – hence the name 'shadow director' – although there is no requirement that they do actually lurk in the shadows or try to remain concealed.

The expression 'shadow director' is a very cinematographic one. It suggests a powerful figure lurking just out of sight. Morritt LJ in *Secretary of State for Trade and Industry v Deverell* (2000) held that the purpose of the rule was to identify people who have a real influence over a substantial amount of the affairs of the company. It was suggested in *Re Hydrodam (Corby) Ltd* (1994) that it must be shown that the directors tended to act on the directions of the shadow director.

The register of directors

It is important in company law that as much transparency as possible is maintained in relation to the company's affairs. Therefore, under s 162 CA 2006 there is a requirement that every company must keep a register of its directors. That register of directors is required to contain particular types of information, such as the name and any former name of the directors, their nationality, their occupation and the country in which that person usually resides. One particularly important provision is that a service address must be provided for each director so that third parties know how to contact directors to serve them with legal proceedings. It had previously been intended that the directors' home addresses might be made available; however, it was decided that a service address through which directors could be contacted would be sufficient and that it would help to preserve the privacy of directors. Indeed, the residential addresses of directors are treated as being protected information (s 240 CA 2006). If a director is itself a corporation, then particulars of that corporation must be provided on the register (s 164 CA 2006). Needless to say, the company must keep the registrar up to date with any changes in the particulars of its directors.

What is a director?

In the companies legislation there are many tasks that are delegated specifically to the directors. In any company's articles of association there are also specific tasks that are delegated to the directors specifically. Generally the board of directors or people who have powers delegated to them by the board of directors are the engine that causes companies to make decisions about the conduct of its business. Consequently, meetings of the board of directors are important in the operation of companies: whether issuing shares, allotting shares, taking decisions as to the business direction of the company, and so forth. Executive directors within any trading company are powerful. Directors of large trading companies usually occupy a particular floor in the company premises, they stride through the building with confident tread and frequently have a designated parking space. In such circumstances, to be a director is to be well paid, respected and powerful. From the perspective of company law, however, to be a director is to carry legal powers and also to bear a wide range of onerous obligations. We considered this paradox of being both powerful and yet being burdened in the previous section.

The composition of the board of directors

Given that the company may be used as a means of operating a cypher for a private individual like Aron Salomon, there have always been rules on how many people must act as directors of the company and so take responsibility for its actions, instead of using a corporate form with only one individual potentially bearing ultimate responsibility for any torts or other wrongs which it may commit.

The rules on the required numbers of directors differ between private and public companies. Under s 154 CA 2006 a private company must have at least one director, whereas a public company must have at least two directors.

The reduction in the possible number of directors of a private company to only one person was a controversial change, because it means that there need be only one person operating the company as a director: the manner in which 'meetings' of directors can be held when there is only one director is an oddity. There were lots of older cases in which it was held to be impossible for there to be a 'meeting' if only one person attended (for example, *Sharp v Dawes* (1876), in which only one person turned up for a meeting but he nevertheless purported to go through the agenda, to take minutes and even to propose a vote of thanks to himself for presiding). Nevertheless, the 2006 Act is concerned to make the operation of companies more straightforward than it was previously.

The reason for public companies having more than one director is that the broader public may be involved in investing in public companies and so there needs to be more than one person responsible for acting as a director so as to minimise the risk of fraud or abuse. It was long a feature of the law relating to unit trusts, for example, that two fiduciaries were required to be involved so that one could ensure that the other one was behaving properly; the same holds true for public companies today.

Under s 155 CA 2006 a company must have at least one director who is a natural person (that is, a human being). Otherwise, it should be recalled that companies may themselves act as directors of other companies because they are legal persons themselves. This nonsense – in which an abstract entity is able to function as a real person – is resolved in practice by the duties of the directorship being imposed on the human directors of whichever company is nominally acting as director. There will of course be many circumstances in which companies may be used to hold shares in other companies or to act nominally as directors of other companies; therefore, it is important that there must be at least one natural person who is a director. The minimum age for a (human) director of a company is 16 (s 157 CA 2006).

The appointment of one director to act as the overall manager of the company is not something that the directors can do of their own initiative at common law. Instead, the articles of association must sanction the appointment of a person as 'managing director' (in the old parlance) or as 'Chief Executive Officer' (in the modern parlance) (for example, *Nelson v James Nelson & Sons* (1914)). The articles will delineate the powers and the means of setting the remuneration of that person.

Preventing directors from excluding their liability

In contract law generally, it is possible for the parties to agree that they will not be legally liable for things that are specified in the contract. However, if company directors were able to exclude their liability (for example in their contract of service), then that would mean that any wrongs which were their fault would not be compensated. Importantly, then, s 232 CA 2006 provides that any provision which purports to exempt the director from any liability, in whole or in part, which would otherwise attached to her in relation to any negligence, default, breach of duty or breach of trust in relation to the company, is void (s 232(1)). Therefore, directors are not able to avoid their fiduciary and other duties simply by some contractual or other provision between them and the company. Similarly, any provision that purports to provide a director with an indemnity against any such liability will also be void (s 232(2)).

Directors are not ordinary employees

It is an oddity of UK company law that directors are not treated as ordinary employees of the company. This is due to the fiduciary nature of a director's office: after all, only directors will always owe fiduciary duties to the company. Other people will only owe fiduciary duties to a company if there was something particular about their relationship with the company that invoked those duties under fiduciary law. Equally significantly, instead of being ordinary employees, directors have service contracts with the company that are necessary to enable directors to be reimbursed for their work on behalf of the company. Otherwise, it would be a breach of the directors' fiduciary duties if they took remuneration from the company to which they owed fiduciary duties. It is necessary that the directors have specific permission to receive such remuneration. (That is also why I have tried to avoid describing directors receiving 'salaries' as though they were ordinary employees, and instead have talked about them receiving 'remuneration' for discharging their office.) As part of the general requirement for transparency in company law, the company is required to keep copies of directors' service contracts available for inspection (s 228 CA 2006). The company's shareholders have a right to request copies of the service contracts and the company must make these service contracts available for inspection by its shareholders.

Transparency in the provision of information is a principal tenet of company law

One of the principal tenets of company law is transparency in the provision of prescribed types of information. Transparency means making as much relevant information about companies available to the public as possible. This enables ordinary investors to decide whether or not to invest in a company, it enables people considering trading with the company to decide whether or not they wish to take the risk of trading with that company, and in general terms it enables all third parties to decide how they want to deal with the company. Therefore, the provision of information is essential both to the operation of commerce and also to the logic of company law. The following sections consider situations in which directors are required to make disclosures.

The obligations on directors to make disclosures

As part of a director's general duties it is important that there are no conflicts of interest between her personal interests and her obligations to the company (s 175 CA 2006). So if there are any potential conflicts of interest, the director must make a disclosure of the matters involved. This is just one example of the sorts of complex decisions which directors will be required to make between competing commercial or legal possible actions. What is more, these sorts of problems may arise in a number of contexts. Consequently, there is a range of provisions in the legislation that requires the directors to make disclosures of a variety of matters, including conflicts of interest, personal involvement in transactions, taking loans from the company, and so forth. This section identifies some of the key provisions that impose these obligations on directors. The common thread running through them is that directors are required to promote transparency so that there is nothing going on behind the scenes that the board of directors, the shareholders or others need to know.

Section 182 CA 2006 requires that a director disclose any interest in a transaction or arrangement that is being entered into by the company. The director must declare the nature and extent of her interest to the other directors. This declaration must be made either at a meeting of the directors or by notice in writing or by means of a general notice. The procedures for these various notices are set out in s 182 and the following sections. If the declaration is inaccurate or incomplete, then the director must make a further declaration. Failure to declare an interest in such a transaction is an offence (s 183 CA 2006).

Section 190 CA 2006 relates to substantial property transactions that require approval of the members. In essence, where a director is to acquire a substantial asset from the company or if the company is to acquire a substantial asset from a director, the company may not enter into that arrangement unless it has been approved by a resolution of the company's members. This principle also applies to transactions involving not just a director but also any person who is connected with such a director. The assets referred to in this provision are specifically non-cash assets. Examples of people who will be connected with the director, as defined in s 252 CA 2006, are members of the director's family and a body corporate with which the director is connected. Clearly it is important to extend this provision to associates of the directors or else it would be too easy to elude these sorts of provisions. The definition of the term 'substantial' transaction in this context is set out in s 191 CA 2006, which provides that an asset is a substantial asset if it exceeds 10 per cent of the company's total asset value or if it exceeds the value of £100,000. These provisions are clearly an important adjunct to the general directors' duties relating to conflicts of interest.

Among the many controls on a director dealing with a company of which she is a director are controls on quasi-loans. So, in s 198 CA 2006, companies are prohibited from making any quasi-loans to a director of the company unless the transaction has been approved by resolution of the company's shareholders. Not content with restricting the circumstances in which directors may receive loans from the company, a quasi-loan is a transaction that in effect involves the company paying for any obligation of the director or reimbursing the director for any personal expense of her own.

The directors' report and the business review

One of the key objectives of company law is to ensure that the public is provided with a wide range of information about the affairs of the company. In theory this provides information for investors, for creditors dealing with the company and for anyone else who may wish to know. The directors of a company are required to prepare a directors' report for each financial year (s 415 CA 2006), and they are also required to approve annual accounts (s 394 CA 2006, as considered in the next section of this chapter). This is an important part of a director's duties and also an important part of providing information to the company's members, and where appropriate to third parties. There is an exemption from this requirement in relation to small companies: in effect this protects directors of small companies from this additional administrative burden. At the general level, the directors' report must give the names of the people who were directors and also set out the principal activities of the company during the course of that financial year.

More importantly perhaps, further to s 417 CA 2006, the directors are required to set out a business review for that company in that financial year. The exception to this requirement is again in relation to small companies. The purpose of the business review is to give information to the company's members so that they can assess the directors' performance in that financial year (s 417(2)). It is noteworthy that this is not intended to be a report for the investing public generally; instead, it is intended to be for the use of the company's members. The business review is required to contain two things: first, 'a fair review of the company's business' and, second, 'a description of the principal risks and uncertainties facing the company' (s 417(3)). That the review must present a 'fair review' of the company's performance, and thereby a review of the directors' performance, means that the report cannot be biased nor can it tendentiously overpraise the directors' performance. In that sense, the review is required to present 'a balanced and comprehensive analysis' of three things: the development of the business, the performance of the business, and ultimately the position of the business at the end of the financial year (s 417(4)).

In relation to a 'quoted company' (that is, a company whose securities are traded on a regulated market), the business review must consider the main trends and factors affecting the future success of the business and also information about the impact of the company's business on the environment, the company's employees, and social and community issues. The review must also consider key performance indicators relevant to the measurement of the success of the particular business in question. For the directors this is as much an art as a science in that there are no strict rules as to precisely what the directors must write in their report: rather they are required to exercise judgment in deciding what is appropriate in the circumstances. In practice the directors will rely on their professional advisors as to the manner in which the report should present the condition of the company as well as the precise contents of that report. In relation to large public companies, such a business review will commonly form part of the accounts and annual report in any event, but will also involve advice from numerous types of professionals and experts, perhaps in the particular business in question. Needless to say, in relation to large public limited companies, such a business review will also be important to prospective (and indeed the current) investors in those companies. The directors' report is also required, further to s 418(2) CA 2006, to contain a statement that, as far as each director is aware, there is no relevant information which is not being provided to the company's auditors and that each director has also taken all of the steps which she ought to have taken to make herself aware of any such information.

The directors of a quoted company must also prepare a report into their remuneration (s 420 CA 2006). It is an important part of corporate governance in the UK, in relation to public companies in particular, that setting the level of directors' remuneration is not left entirely to the directors themselves, but rather is passed on to a specific remuneration committee made up primarily of non-executive directors. (This is discussed in Chapter 9.)

The directors' report is then required to be approved by the board of directors and signed on behalf of the board (s 419(1) CA 2006). The annual accounts and the directors' report must be circulated to every member of the company, to every holder of the company's debentures, and to every person entitled to receive notice of general meetings (s 423 CA 2006). Moreover, a quoted company must ensure that its annual accounts and reports are made available on its website (s 430 CA 2006). In the modern age, the creation of the internet means that information can be more easily and efficiently published online, and company law has consequently seized on the opportunity that it presents. The directors of public companies are required to lay a copy of the company's annual accounts and the directors' report before the company's general meeting (s 437 CA 2006). While perhaps less technologically advanced than the website, this traditional means of publication ensures that the members have all had access to the accounts and to the directors' reports. The obligation to file accounts does not end there, however, because the directors of the company of whatever sort must also deliver copies of the company's accounts and of the reports to the registrar in relation to each financial year.

The Companies Act 2006 contains a number of regimes for reporting by different types and sizes of company. As discussed in Chapter 11, securities regulation requires that public companies make regular financial reports available to the investing public in any event. So, a corporate governance statement is also required under the FSA Disclosure and Transparency Rules in relation to public companies (s 472A CA 2006, as discussed in Chapter 11).

Significantly, further to s 463(2) CA 2006, a director will be liable to compensate the company for any loss that the company suffers as a result of any untrue or misleading statement in a directors' report or as a result of any omission from such a report of anything required to be included in it. Again, this approach is necessitated by the importance of the provision of accurate information by company law.

Accounts and auditors

Accounts

The preparation of corporate accounts is a science that is undertaken by accountants in accordance with the applicable accounting standards. Nevertheless, as the application of those accounting standards in practice often demonstrates, accounting is often practised as an art rather than a science in many circumstances. The significance of having international accounting standards for large multinational companies as well as accounting standards for even the smallest private companies has been established time and again by the frauds and malpractice which have been perpetrated through the misuse of corporate accounts to present a false picture of the company's condition. For example, in the case of Enron there were accounting practices used that treated future losses as though they were current profits when given an optimistic current market value.

Section 393 CA 2006 provides that a company's accounts must 'give a true and fair view' of a company's 'assets, liabilities, financial position and profit or loss' (s 393(1)). It is generally assumed that compliance with ordinary accounting practice will constitute the giving of a 'true and fair view', assuming those accounts have been properly prepared in accordance with ordinary accounting practice. Each company is required to prepare annual accounts (s 394 CA 2006), comprising a profit and loss account and a balance sheet that present a true and fair view of the company (s 396). Section 395 CA 2006 provides that the company may prepare its accounts in accordance with international accounting standards or, as a matter of UK company law, in accordance with s 396. Section 396 CA 2006 in turn provides that a company's accounts must comprise a balance sheet calculated as of the last day of the financial year and a profit and loss account. Both of these documents must give a true and fair view of the company's assets and liabilities, and its profit and loss. The following sections of the Act set out a number of provisions relating to the preparation of individual company accounts and corporate group accounts. The directors should only approve the accounts once they are satisfied that they do present a true and fair view of the company (s 393(1)(a)). In relation to groups of companies, group accounts must be prepared (s 399(2)) unless a specific exemption applies to the group.

The detail of the International Financial Reporting Standards, which are published by the august International Accounting Standards Board, is beyond the scope of this book. However, those are the standards to which accountants cleave when preparing company accounts. The Companies Act 2006 gave a power to the Secretary of State to create regulations for appropriate accounting standards in this jurisdiction. In the UK, CA 2006 differentiates between 'small companies' (with a turnover of not more than £6.5 million, capital assets of not more than £3.26 million, and/or not more than 50 employees), medium-sized companies, public companies and companies involved in financial services business or insurance. Each category has a different accounting standards treatment. There are separate requirements for the publication of financial statements imposed on companies by securities regulation, as discussed in Chapter 11. Typically, annual accounts will include a profit and loss statement and a balance sheet, where the former indicates the profit or loss of the company in a financial year, and the latter indicates the capital assets and liabilities of the company. However, as the Enron scandal and others have shown, it is only cash flow statements that are so difficult to manipulate so that they always present the true condition of the company.

Audit and the liabilities of auditors

The liabilities of auditors

It is a maudlin part of commercial life that companies are too often used as vehicles to commit fraud or become the means through which desperate business practices transform over time into frauds. Given the importance of the accounts in presenting to shareholders and the outside world an understanding of a company's fortunes and prospects, it is important that they are accurate. Consequently, an auditor is required to investigate the accounts and the operation of the business to certify that the accounts present a 'true and fair view' of the company's position. The auditor therefore runs the risk of attracting liability for losses incurred by people who have relied on its certification that the company's accounts are accurate when dealing with the company or investing in the company. A number of significant cases have been generated by litigation of this sort.

The first line of authority relates to the nature of the duty that the auditor owes to the company. The overarching principle is that the auditor's duty is to 'ascertain and state the true financial position of the company at the time of the audit' (*London & General Bank (No 2)* (1895)). In that case it was held by Lindley LJ that it was not the responsibility of the auditors to tell the directors or the shareholders whether or not a dividend could be paid out of the company's capital, merely to certify whether or not the balance sheet presented a true position at the time it was audited. This approach was followed in *Kingston Cotton Mill Co (No 2)* (1896), in which Lopes LJ held that the standard of care owed by an auditor was 'that skill care and caution which a reasonably competent, careful and cautious auditor would use'. The question then arose whether or not an auditor was expected to uncover errors or frauds concealed behind the accounts[.] It was held that an auditor 'is a watchdog, but not a bloodhound', meaning that there was no positive duty on an auditor to find every error or misdemeanour that would not be evident to a reasonably competent auditor exercising reasonable care.

However, since these cases were decided at the end of the 19th century, our expectations of auditors in practice have hardened somewhat. So, in *Re Thomas*

Gerrard & Sons Ltd (1968), Pennycuick J held that the standards against which auditors would be measured were more exacting in 1968 than they were in 1896. In that case, the managing director of a company had caused the accounts to be falsified so as to show greater stock being held by the company than was the case and also suggesting that invoices were payable by the company later than in fact they were (thus worsening the company's cash flow position). It was held that the auditor should have contacted suppliers to find out about the invoices and should have checked the company's stock levels: that was the opinion of an expert witness as to auditing practice, and it was accepted by the court. Therefore, there is an element of an auditor needing to act as more of a bloodhound than was considered to be the case in 1896. Indeed, much of an auditor's role will be governed by the terms of the contract on which the auditor is appointed, but nevertheless auditing practice will tend to govern the level of assiduity that is expected of an auditor. Depending on the type of business that is being audited, it is generally expected that an auditor will enquire into anything that would affect a company's accounts.

The further question is the scope of the duty of care which the auditor owes: specifically, does the auditor owe duties to people other than the company itself, and if so, to whom? The leading decision of the House of Lords in *Caparo Industries plc v Dickman* (1990) concerned a claimant who had acquired shares in Fidelity plc on the basis of an auditor's statement that incorrectly stated the true position of that company. In essence, the opinion of the House of Lords was that an auditor should not be liable in negligence to someone who had relied on a misstatement in its report but who was outside the group of people who could reasonably have been foreseen to rely on that report. This case was therefore of enormous importance in relation to the responsibilities of auditors. A distinction was drawn by Lord Bridge between circumstances in which the maker of a statement was fully aware of the claimant's reliance on it, and circumstances in which a statement (such as an auditor's statement) was being put 'into more or less general circulation and may foreseeably be relied upon by strangers . . . for any one of a variety of purposes which the maker of the statement has no specific reason to anticipate'. The House of Lords was concerned that this would present an auditor with an otherwise open-ended possibility for loss. (The specific nature of this head of liability in relation to securities transactions is considered in detail in Chapter 11.) These principles were confirmed by the House of Lords in *Stone Rolls Ltd v Moore Stephens* (2009).

These common law responsibilities of the auditors should be considered as operating in parallel to the responsibilities set out by the appropriate accounting standards. It should be noted that 'small companies', as defined above, do not require an auditor (s 477 CA 2006).

The statutory code on auditing

The provisions governing audit are set out in Part 16 of the Companies Act. A large number of EU directives and UK implementing measures deal with auditing.

It is in s 475 CA 2006 that the requirement for the company's annual accounts to be audited is set out. Small companies and dormant companies may be exempt from this requirement. Auditors must be appointed in relation to a private company for each financial year unless the directors 'reasonably resolve' that an audit is not required (s 485 CA 2006). The period of appointment of such auditors must be set out (s 487 CA 2006).

In relation to public companies, auditors must be appointed for each financial year unless the directors 'reasonably resolve' that audited accounts are unlikely to be required (s 489 CA 2006): it is suggested that would be a rare circumstance in which this would occur. The significance of audited accounts was brought home by the scandals at Enron, WorldCom and others. In those situations, the senior management of those organisations were able to use their company's attempts to misrepresent the value, profitability and success of their businesses. Properly audited accounts would have burrowed into the background of all of those companies' activities and published financial information to establish whether or not those profits had been genuinely earned. So s 498 CA 2006 provides that the auditor is required to carry out 'such investigations as will enable him to form an opinion' about the company's systems for maintaining accounting records and whether the accounts tally with those records.

In the scandals just mentioned, senior management were either able to pull the wool over the auditors' eyes or the auditors had become so closely linked to the companies that they were auditing that in fact they became bound up in the frauds. Therefore, it is important that auditors are not allowed to become too close to senior management, nor should they be appointed for too long a period so that such a closeness becomes a possibility. The thinking is that while the company's accountants may work closely with the company, the auditors provide an objective, outsider's view of the company, its financial systems and its accounts. Therefore, objectivity and distance is vital. To this end, the auditors are required to prepare a report on the company's annual accounts for all of the company's members (s 495 CA 2006). Importantly, the purpose of the report is to provide information to the company's members (that is, the shareholders), even though this report may in practice be relied upon by other classes of people, such as investors in public markets. (The legal ramifications of such reliance are considered in Chapter 11.) Section 527 CA 2006 provides that the members of a quoted company can require the publication of concerns about the company's audit.

The company secretary

So much of company law is concerned with formal requirements as to the holding of meetings, the maintenance of registers, the lodging of identified documents, and so on, that traditionally someone had to be responsible not only in the sense of being required to carry out those functions but also in the sense of being legally responsible if there was a failure to comply with one of those company law requirements. Therefore, a company secretary was always required to perform

that role. CA 2006 removed the obligation for all companies to have a company secretary. It is now no longer required for a private company to have a company secretary (s 270). The removal of this requirement was a recognition of the fact that it placed a cost burden on small companies to have a company secretary and that, in any event, many small companies used outside professional advisors to advise on many of the tasks performed by a company secretary. However, public companies must still have a company secretary (s 271). A company secretary is someone who is required to have the necessary knowledge and skill to perform that role (s 273(1)), and must either have experience in that role in three out of the five preceding years or be legally or otherwise professionally qualified (s 273(2)).

Significantly, perhaps, the common law's perception of the importance of the company secretary has changed over time. Whereas Lord Esher MR considered the company secretary to be 'a mere servant' whose 'position is that he is to do what he is told', and who could therefore not bind the company due to his lowly rank (*Barnett, Hoares & Co v South London Tramways Co* (1887)), Lord Denning MR took the view that times had changed and that the company secretary 'is a much more important person now than he was in the past. He is the chief administrative officer of the company with extensive duties and responsibilities' (*Panorama Developments (Guildford) Ltd v Fidelis Furnishing Fabrics Ltd* (1971)). Consequently Lord Denning was prepared to hold that a company secretary's acts could bind the company. Nevertheless, that will still depend on the circumstances. Ordinarily, a company secretary would be responsible only for administrative matters such as the organisation of company meetings and other minor matters relating to the company's general activities (perhaps involving hiring cars or employing junior staff), but not ordinarily the negotiation of large-scale corporate financing such as a syndicated loan (*UBAF Ltd v European American Banking Corp* (1984)). Therefore, it will depend on the size and nature of the company and the particular powers of the company secretary as to the extent of her power to bind the organisation. Ordinarily, the company secretary is the bureaucrat within the organisation who ensures that the company's constitution is complied with and that its company law obligations as an entity are performed properly.

Meetings

Shareholder meetings

The role of shareholder meetings

Meetings of shareholders are a key part of what is commonly referred to as 'shareholder democracy'. It is at shareholder meetings that the voting rights held by shareholders whose shares are granted voting rights under the company's constitution can exercise their rights. There are a number of issues, as discussed in various places in this book and as set out by the Companies Act 2006, which require that a given voting majority of the shareholders pass an appropriate resolution: on some occasions this is a 'simple majority' (of more than 50 per cent of the votes) and on other occasions it is a 'special resolution' requiring more than 75 per cent of the votes. (The old 'extraordinary resolution' no longer exists in UK company law.) Consequently, there is an important part of company law that is concerned with the provision of adequate notice of meetings and the process for passing resolutions. We shall discuss some of those principles in this section.

In this section we are concerned with meetings of shareholders such as the annual general meeting (AGM) which used to be required of all companies, but which is now only required for public companies and companies whose articles require it, and other meetings which are called on occasion to consider particularly important issues which are considered to be too urgent to wait for the next AGM. Importantly, though, we are considering meetings of shareholders, not meetings of directors. If we return to the basis of company law, companies were once simply associations of people and it was those people who ultimately owned the company and its assets, and therefore it was for those people to vote on the direction of the company. Therefore, formal meetings of shareholders were essential to deal with important questions within the company, to receive the annual accounts and a report from the directors.

Modern companies are, as we considered in Chapter 3, more complex. The directors now tend to control the management of the company's business and other activities. Nevertheless, company law retains the idea that it is for the shareholders to exercise ultimate control over the affairs of the company, even though the directors are responsible on a day-to-day basis in practice. Therefore, shareholders' meetings are a significant mechanism that shareholders can use to retain and to exercise their control over the directors. Indeed, in the business pages of the newspapers you will occasionally see pictures and reports of harassed executives (that is, the directors) who have faced a stormy meeting of shareholders as they have tried to get the shareholders to vote in favour of a plan which the directors have had for the company. In many private companies, of course, it will be the same human beings who act as directors and as shareholders, but importantly company law considers those human beings to be acting in one capacity when they are acting as directors and in another capacity entirely when they are acting as shareholders. Subject to what is said in Chapter 8 about preventing oppression of one class of shareholders by the others, it is better to think of the same human being as being in law two different people when acting as a director and when acting as a shareholder.

Practical issues relating to voting power

In practice, in large public companies the only shareholders with enough clout to exercise voting majorities are the institutional investors (which are large financial institutions such as pension funds) who collectively own the majority of shares in the largest companies. The management of such companies are usually careful to court the approval of the institutional investors to get them to support significant changes in the business, such as sales of key assets or mergers, long before a meeting is actually held. Therefore, private individual shareholders can usually only exercise any power by gaining publicity for their views in the press and by demonstrating for the benefit of the cameras outside the AGM: that they have such small proportions of the voting rights in large public companies means that they cannot usually win a majority in shareholder votes. Therefore, there is a power dynamic behind shareholder votes, above and beyond the ordinary rules of company law. Nevertheless, companies operate on the basis of democracy. Each voting share will ordinarily carry one vote, and therefore small investors with only a few shares will generally have tiny numbers of votes in public companies.

In relation to private companies, however, things may be different. While public companies will have thousands of investors of various kinds in relation to millions of shares, in a private company it is possible for matters to be much smaller. It is still a requirement of company law, as we shall see, that there are proper notices of meetings and properly organised votes, but the fact that there are fewer share-holders and possibly different power dynamics will make shareholder democracy different. In Chapter 3 we considered the Gilbert family business and we supposed that each family member participating in that business would get a roughly equal number of votes: this would sound to us to be a typically democratic way of conducting business and it would require a genuine majority of the human beings who had put money and effort into that business to agree on the appropriate way forward. However, it is also possible that if a family business was begun by one family member so that she retained for herself 80 per cent of the voting shares, it would mean that that family member would always have a clear majority in votes at shareholder meetings to carry through whatever she wanted to do, regardless of the views of other shareholders. Those other shareholders would have no effective control because their shareholding in this example would amount to only 20 per cent of the votes, even if there were more individuals. (All of this assumes that nothing in this is considered by the court to be oppressive of those other share-holders, as discussed in Chapter 8, 'Shareholders' rights'.)

It is also possible that an outside investor in a family business operated through a private company might insist on being granted enough shares in the company to give her a majority in shareholder votes. If you have ever watched the BBC television programme *Dragon's Den*, you will have seen the investors on that programme ('the dragons') negotiating with entrepreneurs to invest a given amount of money for a 'percentage of the business': this means that the dragons are asking to have sufficient shares transferred or issued to them so that they have a given percentage of the voting rights in the company. The entrepreneur is always reluctant to give away control of their company in spite of the amount of money involved, because effectively this gives the dragon with a majority shareholding the power to dismiss directors, to approve reorganisations of the company, and even to change the company's constitution with the right majority.

As was explained in Chapter 5, there are different classes of shares in many companies. Therefore, it is also common in companies with different classes of shares to have meetings of shareholders in a particular class so that they can consider issues of interest in common to them. Ordinary shares will usually carry one vote per share. Some preference shares do not have any voting rights at all, but rather have rights to a fixed dividend – which means that the other shareholders have control of the company but the company is required to pay out a fixed amount annually to the preferred shareholders. It is also possible that particular classes of shares will be provided to have huge numbers of votes attached to them, so that the owner of those shares retains effective control of the company – this would be more likely in a private company because it is unlikely that institutional investors would invest in a public company on that basis. In relation to takeovers, for example, it is possible to insert a 'poison pill' in a company's constitution so as to deter takeovers by providing that a given class of shares would lose all of its voting rights on the happening of a particular event (for example, the making of a takeover offer) or that another class of shares would have its voting rights increased. In effect, the purchaser would have no control over the company in those circumstances.

The mechanics of shareholder meetings – calling meetings

The directors have the power to call general meetings in accordance with the terms of the articles of association (s 302 CA 2006). If a sufficient number of shareholders (in different contexts) demand a general meeting, the directors will be compelled to call one (ss 303–305). The court may also demand the calling of such a meeting in appropriate circumstances (s 306).

Further to s 307, members must be given at least 21 days' notice of an AGM, and 14 days' notice for all other meetings, unless the company's articles of association require a longer notice period. A shorter period can only be used if a 95 per cent majority of shareholders in a public company agree to it (and 90 per cent in a private company), provided that they have notice of the resolutions that are to be considered at that meeting. The notice periods clearly give the shareholders time to consider the issues, to gather information and to do mundane things like making travel plans.

The notice must be sent to all members and directors of the company (including their personal representatives if they have died) (s 310), and the auditors also have a right to receive a notice (s 502). The notice must state the time, date and place of the meeting and the general nature of the business that is to be conducted at the meeting (s 311), and it must comply with any other requirements in the articles of association. The concern underlying these provisions is that a meeting might be organised which fails to describe its business properly, so that, for example, the directors are able to change the direction of the company without all of the shareholders having thought it worthwhile to attend the meeting. It has long been the position in English law that a notice must 'give at any rate a fair, candid and reasonable explanation' of the business which is proposed to be discussed at the meeting (*Kaye v Croydon Tramways* (1898)).

A 'special notice' procedure set out in s 312 CA 2006 is required in some circumstances, for example where the auditor is to be changed (s 510 and s 514) or there is a resolution to remove a director from the company (further to s 168). In essence, this also requires advertisement of the meeting in national newspapers. Not infrequently such small box advertisements appear in the *Financial Times* in particular, announcing that a specified item of business is to be conducted at a meeting of a given company. Accidental failures to give notice of either a resolution or of a meeting may now be excused by s 313 CA 2006, except where the articles provide to the contrary, in some of the circumstances just considered.

The mechanics of shareholder meetings – the conduct of the meeting

The Companies Act 2006 contains a number of provisions relating to the proper conduct of meetings. Before coming to those principles, however, it is important to recall that the precise rules governing the operation of meetings and the matters which need to be approved by the shareholders are governed by the articles of association of each company, except to the extent that mandatory rules of the CA 2006 do not permit any other method to be used.

General meetings of shareholders must have a quorum of appropriate people present. This means that there must be a minimum number of the right people present so that any vote taken at the meeting can be considered to be valid. The appropriate quorum will be identified in the company's articles of association, as is required by Article 38 of the Model Articles for example. The appropriate quorum for each different kind of company is set out in s 318 CA 2006 in general terms: ordinarily two qualifying shareholders would be a sufficient quorum, unless the company is a one-member company (in which case one member would obviously suffice). That provision also explains the various permutations when the only people present are proxies of shareholders or other representatives of members of the company. In theory any member of the company could be elected to chair the meeting, although the articles of association would usually provide that a specific person will act – usually the chairman of the board of directors.

The chair is then ordinarily required by the common law to preserve order at the meeting, to ensure that proceedings are conducted properly (including votes at the meeting), to deal with any procedural and other questions which arise during the meeting, and more generally to 'take the sense of the meeting' (which means understanding what the general view of those present is in relation to particular questions without necessarily needing a particular vote, and allowing people to contribute appropriately). An example of the sorts of questions which might arise is *Byng v London Life Association Ltd* (1990), where those attending the meeting had to sit in different rooms because there was insufficient space, which was considered appropriate in the circumstances on the basis that the various rooms were connected by video link and it was possible to hear what was being said in each room. If the links could not fulfil these functions, then the chair of the meeting would have to decide how best to proceed with the business of the meeting.

The mechanics of shareholder meetings – voting

At common law, voting at a shareholder meeting could take place simply by a show of hands (for example, *Re Horbury Bridge Coal Co* (1879)). It is generally taken to be the case that everyone present has one vote, although an objection may of course be raised that the voting power of some shareholders is greater than others due to their shareholding. The chair is ordinarily empowered to deal with any objections further to Art 43 of the Model Articles (in relation to public companies). Under s 320 CA 2006 a record of the number of voting shares with votes cast in support of a motion or against a motion may be taken, rather than relying simply on the show of hands. When an objection to a vote by show of hands is sustained by the chair, or when the chair anticipates that the question to be voted on is of sufficient importance to require it, then a poll of the shareholders and their precise number of votes is taken. It is suggested that except for the most insubstantial questions in relation to large companies, a poll would always be the better way to proceed because it is only by a poll that the number of voting shares can be ascertained. (This is another example of the quaint approach of company common law to some of these questions.)

As mentioned above, the CA 2006 provides for a number of mandatory rules that do not permit the articles of association to provide for an alternative method of conducting a meeting, although otherwise the articles will take priority. An example of these sorts of mandatory rules is contained in Part 13 of the CA 2006, which provides for the means by which shareholder resolutions can be passed. Section 281(1) CA 2006 provides that in a private company a resolution may be passed either in the form of a written resolution circulated before a meeting or at a meeting itself, whereas s 281(2) provides that in relation to public companies such a resolution may only be passed at a general meeting or at a meeting of a given class of shareholders. It is common in the CA 2006 for the rules relating to private companies to be more permissive than the rules in relation to public companies because it is more convenient and cost-effective for a small, private company to be able to proceed with lesser bureaucracy, whereas public companies have such a diverse range of share-holders drawn from the public that matters need to be more formalised.

Nevertheless, s 281(4) preserves the common law principle that taking an informal assessment of the shareholders will be sufficient in many circumstances (*Re Duomatic* (1969)), as with the discussion of voting by show of hands above. The *Duomatic* principle importantly rests on the agreement of the members of the company to proceed informally with some of the company's business, and there-fore is considered by the courts to be a useful mechanism for companies to conduct parts of their business without undue bureaucracy. It is suggested that this is appropriate only for companies which have small enough numbers of shareholders to ensure that no individual shareholder would be disadvantaged by such a procedure: after all, the statutory mechanisms exist to ensure that meetings are arranged and conducted properly in the interests of everyone.

In general terms, as provided by s 281(3) CA 2006, it is an ordinary resolution of the shareholders that is required: that is, a simple majority of the votes cast (s 282). However, where a specific provision of the CA 2006 requires a different form of resolution, then that is the type of resolution that will be required in that context. Typically, that will be a 'special resolution' that requires the approval of more than 75 per cent of the votes cast (s 283).

There are circumstances in which shareholders will not be able to attend a meeting in person. Therefore, it is provided that shareholders will be given the option of voting by proxy instead of attending in person. Shareholders have a right to vote by proxy further to s 324(1) CA 2006. Notice of a meeting must state that the member has a right to vote by proxy (s 325(1)). When the notice of a meeting is circulated, a proxy form (including boxes to vote for or against the resolutions that are to be debated at the meeting) is enclosed. Section 285 CA 2006 governs the process of proxy voting.

Section 287 CA 2006 confirms that any company's articles of association may provide for the admissibility of a vote and for the circumstances in which a vote may be challenged. Usually the ruling of the chairman will be decisive. Nevertheless, that will not prevent a legal challenge being brought subsequently.

Directors' meetings

The meetings of the board of directors, as opposed to the general meetings of the company discussed above where shareholders vote qua shareholders, are again set out most significantly in a company's articles of association. The culture of any given company's directors' meetings will be a matter for that company. Some directors may in practice be dominant; some may often absent themselves from meetings; others may simply be unable to attend all of the meetings because they are in another country or are unwell. Most of the legal rules relating to the conduct of directors' meetings were created by the common law. So, if the articles of association provide that not all of the directors need attend meetings of the board, those articles may provide for the number that will constitute a quorum, or else a majority of the directors must be present (for example, *York Tramways Co v Willows* (1882)). Similarly, votes may be carried with a majority of the directors who are present at the meeting (*Lyster's Case* (1867)), unless the articles provide otherwise. All of the directors should receive notice of the meeting, again in the manner stipulated by the articles of association. It is noticeable that many of the cases that established these rules predate *Salomon v A Salomon & Co Ltd* (1897) and therefore we might recognise that in a world of email, teleconferencing and communication almost anywhere by hand-held electronic devices, that the organisation of such meetings might be more pragmatic in the modern age.

Nevertheless, the proper conduct of meetings in accordance with the corporate governance principles, which are described in Chapter 9, so that all views can be aired remains important; particularly in the wake of the wave of corporate scandals which hit very large companies in the UK and the USA at the turn of the millennium, where senior executives had mismanaged their organisations criminally without anyone being able suitably to question or challenge their actions. Directors have rights, such as the right to inspect the company's books and accounts (for example, *Burn v London and South Wales Coal Co* (1890)), which are intended to empower directors to challenge the actions of the company. In practice, however, other powers, such as the power to delegate the powers of the directors to subcommittees (if authorised to do so by the articles: *Howard's Case* (1886)), both enable groups of executives to segregate control of particular aspects of the business to themselves on the downside and also empower non-executive directors to take control of sensitive topics such as the remuneration of the executive directors on the upside. As we have noted, there is nothing necessarily good or bad about the company structure: instead much turns on the way in which the company is operated in practice. Importantly, s 280 CA 2006 provides that nothing which the Act requires to be done by a director or by the directors together can be delegated to someone else precisely to prevent this sort of abuse.

One issue that clearly arises is in relation to disagreements between the shareholders in general meetings and the directors in meetings of the board of directors: whose view will be decisive? Whether the general meeting of shareholders can instruct the directors to act in a particular way will depend upon the terms of the articles of association of the company in question (for example, *Salomon v Quin & Axtens Ltd* (1906)). The courts have generally taken the approach that the decisions of the directors as to the proper management of the business will not be capable of interference unless permitted by the articles (or as discussed in Chapter 8 where there is oppression of the shareholders); and the courts will certainly not substitute their own view of what the company should have done for the decision of the directors who actually manage the business (for example, *Automatic Self-Cleansing Filter Co v Cunninghame* (1906)). This approach does reflect the de facto organisation of power within modern companies, in which the directors manage the business and the shareholders are reduced to the status of mere investors hoping that there are sufficient profits to be distributed among them.

The operation of some corporate group meetings in practice

One of the oddities of the *Salomon* principle is that each subsidiary company within a corporate group is a distinct company, and therefore all of the formalities for its operation (meetings, resolutions, decisions by the board of directors in the best interests of that company) must be performed, even if it has been created simply as an administrative unit for some larger purpose. Therefore, each subsidiary needs to go through the motions of having these companies operated as though they are distinct entities in reality. (This may be necessary in case a subsidiary is sold off at some point in the future, or else the new owners may sue the previous owners in the name of the company for failing to operate it properly.)

It is worth spending a moment thinking about how corporate groups work in this way. The senior executives within a corporate group will sit as a board of directors of the holding company and as far as the ordinary employees are concerned, they will simply run 'the business', with particular people in charge of different departments or divisions within that company. (At that level, it is a little like an army, with various ranks and different divisions taking on different responsibilities.) Those senior executives will focus on the commercial aspects of the business without necessarily thinking much about the legal niceties until their lawyers point out that there is an issue. Typically there will be a person who is company secretary for that corporate group. That person will be responsible for ensuring that each subsidiary company maintains records and holds meetings, which at least nominally treat that subsidiary as a distinct legal person in compliance with company law, even though commercially the executives may be little concerned about how matters are distributed across those subsidiaries on a day-to-day basis. Within corporate groups it is often just a mundane part of an employee's job to sit as a director of a subsidiary company. The holding company will own all of the shares in these subsidiary companies, and therefore delegates from the holding company will be required to vote at a general meeting of that subsidiary company (unless the company elects not to have a general meeting, as discussed above).

Generally this is a tedious experience because the directors' meetings of these subsidiary companies will usually be held back-to-back across a long afternoon. The executives will already have decided which assets will be moved between companies and what the ultimate goals of each subsidiary are. Then the company secretary will organise the paperwork to be distributed to all of the people who sit as directors of these subsidiary companies (in accordance with the procedures discussed in Chapter 6): this paperwork will usually be distributed in the internal mail and an email circulated the day before the meeting to remind everyone that they are needed to attend the meeting. The minutes of the meeting are likely to have been prepared in advance and then the directors sit for a number of hours as the company secretary prompts the nominal chairmen of those subsidiaries to ratify what has been agreed in advance. Occasionally, some of the people in that meeting room will leave and be replaced by others. Fresh coffee and more biscuits will be wheeled in as one company's business after another is sorted out. At this level, company law becomes a purely administrative activity. For those people who ask nothing more from their working life than a warm office and gentle, undemanding administrative job which requires a little technical legal knowledge, the life of a company secretary in a large group of companies may be ideal. And there are biscuits too!

Moving on ...

This long chapter has attempted to highlight some of the key aspects of company law in relation to the operation of a company in practice. In the latter stages of this chapter we have begun to identify some of the problems that might arise when there are disagreements among shareholders as to the future direction of the company, or similar disagreements among directors, or disagreements between the directors and the shareholders. These issues raise questions of 'corporate governance' which impact on the duties of the directors to the company, the rights of shareholders and the manner in which a company should make decisions in practice in the best interests of all of its stakeholders. These questions are considered in turn in the three chapters to follow. They are the heart of company law, and raise some of its most interesting questions. We shall begin with an analysis of the general duties of directors.

Chapter 7

Directors' duties

Introduction

So far in this book we have considered the history of company law and the legal foundations of the modern company, and the previous chapter explained how a company is operated. In that chapter we saw how significant the directors and other officers are in the operation of any company, whether large or small. The key obligations for managing the company are placed in the hands of the directors. The directors owe fiduciary duties to the company and a number of common law duties besides. In this chapter we shall analyse the general duties that are imposed on directors to govern all of their activities when dealing with the company. This is the beating heart of company law. The duties that the directors owe to the company, and the effect that that can have on shareholders, employees and others, is central to the functioning of any company. Company law as a discipline is about regulating the way in which those fiduciaries interact with the other stakeholders in the company.

This chapter deals with the general duties of company directors as set out in statute and in the case law. This does not relate to the specific powers or duties considered in the previous chapter or elsewhere in this book. Nor does it consider specific provisions in the articles of association of any given company: it would be possible to provide a power or an obligation for a director of any given company to do or not do a given act in the articles. Here we are concerned with the general duties of directors under UK company law.

An introduction to the statutory code on directors' duties

The purpose of the codification

The law on the duties of directors is, in my opinion, the most interesting part of company law. If I am honest with you, it is the reason I wanted to write about company law in the first place. Those duties were codified in statutory form for the first time in 2006. This new code of duties was imposed specifically on company directors by s 170 *et seq.* of the Companies Act 2006 (CA 2006).

The objective underlying the creation of this code was that it would make it possible for ordinary company directors with no legal experience to understand their duties more easily. Previously, the huge amount of case law in this area could make it difficult for many company directors to know what their obligations were. Therefore, what the CA 2006 does is to attempt to codify the previous common law and equitable principles in an easy-to-understand statutory code.

The continuing importance of the case law on directors' duties

It is true to say that company directors are now governed by a statutory code that is specific to them. However, one of the most significant features of that code is that it bases itself explicitly on the previous case law that set out directors' duties. Section 170(3) CA 2006 provides that:

> The general duties are based on certain common law rules and equitable principles as they apply in relation to directors and have effect in place of those rules and principles as regards the duties owed to a company by a director.

The statutory code is therefore 'based on' the common law and equitable principles. Significantly, though, the statutory principles have effect 'in place of' the case law. It is suggested that this means that the statute replaces any errant *dicta* in the case law that appear to contradict the statutory provisions; nevertheless, the statutory principles are to be interpreted in accordance with the earlier case law because, after all, the statutory provisions were intended primarily to codify that case law. Section 170(4) CA 2006 provides that:

> The general duties shall be interpreted and applied in the same way as common law rules or equitable principles, and regard shall be had to the corresponding common law rules and equitable principles in interpreting and applying the general duties.

Therefore, while the statutory rules act 'in place of' the pre-existing case law, the statutory rules are nevertheless to be interpreted as though they were themselves case law principles. That means that the high-level principles set out in ss 171 to 177 CA 2006 are capable of being developed by subsequent case law. It is suggested that the case law should not overrule nor materially alter the statutory rules in the way that the courts might otherwise do with case law precedent.

Directors are fiduciaries

Before coming to the particular duties identified in the statutory code, it is important to understand one of the most significant underlying principles relating to directors' duties: company directors owe fiduciary duties to the company. As Lord Cranworth LC held in *Aberdeen Railway Co v Blaikie Bros* (1854), 471:

> The Directors are a body to whom is delegated the duty of managing the general affairs of the Company. A corporate body can only act by agents, and it is of course the duty of those agents so to act as best to promote the interests of the corporation whose affairs they are conducting. Such agents have duties to discharge of a fiduciary nature towards their principal. And it is a rule of universal application that no one, having such duties to discharge, shall be allowed to enter into engagements in which he has, or can have, a personal interest conflicting, or which possibly may conflict, with the interests of those whom he is bound to protect.

Consequently, the directors will be personally liable to account to the company for any breach of their fiduciary duties (*Selangor v Cradock (No3)* (1968)), and will be liable to hold any unauthorised profits on constructive trust for the company as a result of their breach of duties (*Regal v Gulliver* (1942)). As will emerge, the director also owes the usual fiduciary duties to act selflessly in the interests of her principal (in this case, the company).

There are a number of issues remaining in the case law that are not disposed of by the statutory code. In the light of s 170 CA 2006, considered above, the general case law principles, such as the general law on fiduciary duties, will continue to apply to directors. In examining that case law, we shall observe some undercurrents in some of the cases that are specific to company law – especially the idea that directors may be allowed to permit conflicts of interest in some circumstances where the old case law was considered to be too strict. So, when examining this statutory code, we have to examine the case law that preceded it, because both constitute the law on directors' duties from now on.

To whom are a director's duties owed?

It is provided in s 170 CA 2006 that a director owes her duties directly to the company (as opposed to the shareholders or some other person) in the following terms:

> (1) The general duties specified in sections 171 to 177 are owed by a director of a company to the company.

Consequently, if there is any litigation to be brought in relation to a breach of a director's duty, then it is the company that must bring that litigation, because those duties are owed to the company. Moreover, s 170(5) CA 2006 provides that '[t]he general duties apply to shadow directors where, and to the extent that, the corresponding common law rules or equitable principles so apply'.

The way in which the statutory code is discussed in this chapter

We shall take each of the general statutory duties in turn. Because the statutory provisions now act 'in place of' the case law, we shall consider the statutory language governing each duty first; but because the statute is to be analysed in the light of the case law, we shall analyse the case law second and consider how it interacts with the statute.

All references in this chapter are to the Companies Act 2006, unless otherwise stated.

The duty to act within powers

The statutory principle

The first duty of directors is that they must act within the terms of their powers as set out in the company's articles of association and any memorandum of association (that is, its constitution). A company is a person who must act not only in compliance with the general law but also in accordance with the terms of its own constitution. Therefore, the directors cannot act beyond either the terms of the company's constitution or beyond the terms of their own powers within that constitution. Consequently, s 171 CA 2006 provides that:

> A director of a company must –
>
> (a) act in accordance with the company's constitution, and
> (b) only exercise powers for the purposes for which they are conferred.

Under the case law there have been a number of situations in which, for example, directors have used a power to issue new shares, which was intended to enable the company to raise capital, so as to frustrate a takeover attempt without asking the shareholders whether or not they wanted that takeover to go ahead. Consequently, a power intended for one purpose was being used for a different, inappropriate purpose. In *Re Smith and Fawcett Ltd* (1942) it was held that the directors were required to act 'bona fide in that they consider – not what a court may consider – is in the interests of the company, and not for any collateral purpose' (per Lord Greene, 306). If a director breaches any of the obligations under s 171, then that director will be liable to compensate the company for any loss suffered by the company as a result.

The principle in the case law

As was set out above, in *Re Smith and Fawcett Ltd* it was held that the directors were required to act 'bona fide in that they consider – not what a court may consider – is in the interests of the company, and not for any collateral purpose'. This common law principle has given birth to two statutory principles: the principle that directors must act in accordance with the company's constitution and their powers, and also the principle that directors must promote the success of the company (under s 172, considered next). In this context, the directors must be exercising their powers properly and in an honest belief that they are being used in the interests of the company.

A difficult question arises when the directors use a power that has been given to them perfectly properly but for a purpose that is arguably contrary to the objective that that power was supposed to have. One of the most common ways (in the decided case law) in which directors have tended to abuse their powers in this way has been to obstruct takeovers of the company that they considered to be undesirable. So in *Punt v Symons* (1903), directors were found to have used their power to issue new shares for the purposes of issuing shares to people who would agree to their plan of blocking a proposed takeover that they considered to be undesirable. Byrne J held that the power to issue shares had been given to the directors 'for the purpose of enabling them to raise capital when required for the purposes of the company'; but his Lordship considered that instead of using the power to raise capital, the directors had engaged in 'a limited issue of shares to persons who are obviously meant and intended to secure the necessary statutory majority in a particular interest to prevent a vote in favour of that takeover'. Consequently, Byrne J found that 'I do not think that is a fair and bona fide exercise of the power' (at 515).

Similarly, in *Hogg v Cramphorn* (1967), Colonel Cramphorn was a director of a company who dominated the board of directors in practice. Cramphorn wanted to stop Baxter from taking the company over. To do this, Cramphorn cajoled the other directors into issuing shares to people who would vote against the takeover. The question arose whether or not this was a breach of the director's powers to issue shares. It was held that the power to issue share capital was a fiduciary power which could be set aside if it was exercised for an improper motive. Cramphorn had argued that he genuinely believed that what had been done had been in the interests of the company. Nevertheless, it was held further that such an issue of shares could be set aside, even if it had been made in good faith in the belief that it was in the interests of the company. Therefore, the use of the power was held to be beyond the powers of the directors because it had been done improperly. As an interesting postscript to this case, however, this ultra vires exercise of the directors' powers was ratifiable by the shareholders at a general meeting, and indeed the shareholders did vote to ratify it. Therefore, the issue of shares to block the takeover was ratified by the shareholders, so that the exercise of the power became enforceable as a result.

Significantly, s 171(b) requires directors actively to act in accordance with the company's constitution and to operate their powers properly; unlike the approach in *Smith and Fawcett*, which requires that directors simply refrain from committing a breach of their duties.

Importantly the court will not interpose its own view of what the director ought to have done. Instead, Jonathan Parker J explained in *Regentcrest Ltd v Cohen*

(2001) that the court requires that the director make the decision in accordance with an honest belief that it was in the interests of the company (an idea which is considered in relation to the second general duty below). If that decision is detrimental to the company, however, his Lordship held that it would be more difficult to convince the court that such an honest belief existed.

Of course, it is not enough that a director was honest, but rather (as discussed above) that directors must have been exercising their powers for their proper purpose. Self-evidently, the question will then be whether or not the director is acting honestly and also whether or not the application of the director's power was in the best interests of the company. So, in cases like the decision of the Privy Council in *Howard Smith Ltd v Ampol Petroleum Ltd* (1974), the directors of a company purported to be acting in the best interests of the company when issuing enough new shares to cancel out the previous majority shareholding, so that they could garner enough shareholder votes to block a takeover which they considered to be undesirable for the company. It was held, in effect, that an honest belief that the course of action was in the best interests of the company was not sufficient in itself if the directors' exercise of their powers was an inappropriate exercise of those powers. Here, it was held that the proper use of a power to issue shares was to raise capital for the company, and not to block a takeover bid. (After all, the shareholders are permitted to vote for whatever they want in accordance with their shareholding, no matter what the directors may think.) The Privy Council did not replace the directors' judgment of what was in the interests of the company with its own view. Instead, the court focused on the appropriateness of the use of the power in that context.

Therefore, the court will have to consider the purpose for which the power was conferred, and then whether or not the directors exercised that power properly. It is also required that the directors had an honest belief that what they were doing was in the best interests of the company, as is considered next. If the directors simply breach a provision of the company's constitution as to their powers and responsibilities, then they will be in breach of duty.

The duty to promote the success of the company

The statutory principle

The concept of the success of the company

The CA 2006 created a new principle that the directors should for the first time be obliged 'to promote the success of the company'. Some commentators had argued that this principle should always have been part of company law. The difficulty was in knowing what would constitute the 'success' of the company: should that be understood in terms of the long-term profitability of the company, the short-term profits which could be distributed to shareholders, the happiness of employees, the happiness of customers, or something else? The *Company Law Review*, which laid the ground for the CA 2006, explained that:

what is in view is not the individual interests of members, but their interests as members of an association with the purposes and the mutual arrangements embodied in the constitution; the objective is to be achieved by the directors successfully managing the complex of relationships and resources which comprise the company's undertaking.

This policy recognises that there will be a complex web of demands being made on the directors of a company and prioritises the interests of the shareholders ('members') as a group (as opposed to the interests of any particular shareholders). The principle itself is set out in s 172 CA 2006 by identifying six factors which the directors are required to take into account when promoting the best interests of the company. What emerges is a broad range of factors that the directors must consider. Section 172 of the CA 2006 provides that:

(1) A director of a company must act in the way he considers, in good faith, would be most likely to promote the success of the company for the benefit of its members as a whole, and in doing so have regard (among other matters) to –

(a) the likely consequences of any decision in the long term,

(b) the interests of the company's employees,

(c) the need to foster the company's business relationships with suppliers, customers and others,

(d) the impact of the company's operations on the community and the environment,

(e) the desirability of the company maintaining a reputation for high standards of business conduct, and

(f) the need to act fairly as between members of the company.

Clearly, this duty is very broad and it is not focused exclusively on the profitability of the company. Instead it takes into account the company's place in the community and in its interactions with third parties other than the shareholders. It is an understanding of the role and place of companies in society that is much more in tune with the 21st-century understanding of the obligations of business. This is a significant change from earlier understanding of the role of the company director, especially in the USA where in *Dodge v Ford* the Michigan Supreme Court took the view that the directors should focus solely on earning profit for the company and not use the company's money for any other purpose. On such an understanding of company law, many of the six factors which are set out in s 172 would be considered inappropriate. The position under English company law was always slightly different in that in *Hutton v West Cork Railway* (1883) it was held that the company may play a social role, with the result that (as the judge put it) there may be 'cakes and ale' to keep the employees happy and so all of the company's profits do not need to be distributed to the shareholders. We shall consider the six factors that directors are required to bear in mind one at a time.

*Important features of the six factors which directors are required
to consider*

There are three important aspects to the six factors listed in s 172(1). First,
s 172(1) *requires* that directors consider the six factors. It is important that this is
obligatory for directors, and not something that they can choose to do or not do.
The subsection opens with the verb 'must' and ends by saying that 'in doing so'
regard is to be had to the six factors. Therefore, it is suggested that the verb 'must'
qualifies the entire provision and so requires that the six factors be considered. It
may be that, after mature reflection, the directors decide that some or all of the
factors do not require a change in their decisions, but nevertheless the directors
are required to consider them.

Second, the six factors in s 172(1) are not an exclusive list. Instead, s 172(1)
makes it clear that these factors are to be considered 'among other' factors.
Therefore, the directors must identify what they consider is in the best interests of
the company. To this end, s 172(1) provides: 'a director of a company must act in
a way which he considers . . . would be most likely to promote the success of the
company'. As in *Regentcrest Ltd v Cohen* (2001), the courts will not impose their
own view of what the best interests of the company are: that is entirely a matter
for the directors. But the directors would have to be able to justify *ex post facto*
any decision that replaced the six statutory factors with other considerations.

Third, there is a requirement that the directors must be acting in good faith
when they contend that any given course of action will promote the success of the
company.

The six factors

We shall consider each of the six factors in turn, and consider what they might
mean for directors in the future. One theme that emerges is that the CA 2006 was
created with the interests of the broader economy in mind, and not simply with the
interests of the participants in a company (directors, shareholders and so forth) at
the forefront. Instead, the interests of employees, of creditors, and of society and
environment more broadly are brought within the ambit of things that directors
must consider.

(1) Directors must have regard to 'the likely consequences of any decision in the
     long term'. It should not be forgotten that this Act was passed during the
     tenure of a Labour government and that leftist commentators had always
     complained about the overly short-term focus of too much commercial
     activity, of financial markets and of commercial law exclusively on profit.
     This militates against the needs of the economy more generally, which are for
     stable economic activity and, it is assumed, growth. Nevertheless, to ensure
     easy access to capital on financial markets, directors of large public compa-
     nies generally need to show profit growth year-on-year. Therefore, having to

consider 'the likely consequences of any decision in the long term' bolsters the need to look to the larger context of corporate actions over time.

(2) Directors must have regard to 'the interests of the company's employees'. For the first time in a company law statute, employees acquire a right at the very least to have their interests considered by directors when making decisions. Even though in most circumstances management would consider the need for good relations with workers among the other factors that influence a company's good health, this provision creates a *duty* on directors to consider their position. It also means that directors may not be criticised by shareholders or creditors for giving attention to the needs of employees.

(3) Directors must have regard to 'the need to foster the company's business relationships with suppliers, customers and others'. A successful economy requires that all participants in the economic chain – whether suppliers, customers or service providers – interact successfully. So, requiring directors to consider other economic actors is hoped to have a beneficial effect on the economy more generally. Section 172(3) of the CA 2006 provides that the interests of creditors should also be considered.

(4) Directors must have regard to 'the impact of the company's operations on the community and the environment'. Whereas companies might once have sought profit without regard for the environment (for example, with the construction of carbon-intensive industries during the Industrial Revolution) or without regard for the impact on the community more generally (requiring the introduction of consumer legislation, anti-pollution legislation, and so forth), this provision empowers directors to consider those sorts of questions, as well as requiring them to do so. These sorts of considerations are analysed in Chapter 14 'Corporate social responsibility'.

(5) Directors must have regard to 'the desirability of the company maintaining a reputation for high standards of business conduct'. In this provision the Act also presents an ethical project to raise standards of behaviour among commercial people. It is a little mealy-mouthed as drafted, though: one might have hoped that the directors did not need to give a moment's thought to whether or not their company should maintain a reputation for high standards in business conduct. We might have hoped that that much would have been obvious. Perhaps it is a comment on our commercial culture that thinking about whether or not this is even desirable is something which needs to be enshrined in law. A better approach might have been to legislate for a code of good behaviour for directors. In the absence of such a code of conduct, the general duties of directors will have to suffice.

(6) Directors must have regard to 'the need to act fairly as between members of the company'. To ensure the equal treatment of all shareholders is central to company law (except, of course, for the enshrining of the logic of majority rule in *Foss v Harbottle,* as discussed in Chapter 8). The balancing of the rights of different shareholders is considered in the next chapter.

The principle in the case law

There was an understanding in the case law that directors should act in the best interests of the company, but it did not go as far as to identify a consensus that directors were obliged to promote the success of the company, and it certainly did not set out equivalents to all six of the factors considered above. Nevertheless, cases such as *Punt v Symons* (1903), *Hogg v Cramphorn* (1967) and *Howard Smith v Ampol* (1974), which were considered in the previous section, held that the directors must both use their powers properly and also that they must act in the best interests of the company. Nevertheless, those case law *dicta* do not delve as far into this principle as s 172 indicates.

Interpreting the 'success of the company' where there are different interests

One of the ways in which the case law has conceived of the best interests of the company has been by establishing what is *not* in the best interests of the company – as with the takeover cases considered above. Many of the difficulties arise when competing shareholders or competing groups of directors with different views on the best future for the company clash. For example, in a family company some shareholders may want the company's profits to be used immediately to pay them a large dividend, whereas the family members may prefer to have those profits invested back into the company. The best interests of the members as a collective may be difficult to identify, as may the best interests of the company as a business. What is clear from cases like *Mills v Mills* (1938) is that the directors cannot choose arbitrarily to prefer the views of one group of shareholders over another, nor can the directors choose to bind themselves to follow the views of someone who nominated them to the board without considering the best interests of the company, as in *Re Neath Rugby Club Ltd* (2009).

As a result, the idea of the best interests of the company is generally explained negatively in the cases on the basis of what is not an acceptable act in the best interests of the company. So, in *Item Software (UK) Ltd v Fassihi* (2004), a director wanted to divert business from the company to himself and he therefore encouraged the company's suppliers to make their terms so stringent that they would not be acceptable to the company. This director's conduct led to negotiations between the supplier and the company breaking down. The director was clearly not acting in the best interests of the company. However, such an extreme case offers us little indication of what will constitute the best interests of a company.

The duty to exercise independent judgment

Directors are responsible for their own decisions and so cannot bind themselves mechanically by claiming to act in accordance with the wishes of other people. In this vein, s 173(1) CA 2006 provides that:

A director of a company must exercise independent judgment.

However, it is not only direction from a third person that is considered (such as the person who ultimately controls a shareholding in the company, or who employs or retains the director in question), but rather it also encompasses senior executives or shadow directors in the company who exercise a de facto power over the directors. A large number of corporate scandals have involved control of directors by one particular, dominant director or by a small group of powerful directors. So, in *Re City Equitable Fire Insurance Co. Ltd* (1925), there was one senior executive, Bevan, who controlled the other directors so that Bevan was able to commit numerous frauds with the company's property; and the collapse of the Mirror group of companies revealed that Robert Maxwell had the same control over his fellow directors. In both the Enron and WorldCom collapses there was dominance by a small group of senior executives, which meant that the company's financial reports and business activities could be falsified. It is a feature of many human organisations that one or two people may exercise de facto control far in excess of their legal powers. Therefore, directors also need to exercise independence in the sense that they resist control by other people. A fiduciary owes it to their principal to think for themselves.

Section 173(2) provides that there is no breach of this duty if a director acts either in accordance with an agreement entered into properly with the company, or if the company's constitution authorises the directors to act in a particular way. However, the case law had provided that a director could not fetter her own discretion in advance by committing to a particular course of action no matter what, for example by agreeing to support a colleague on the board of directors in all circumstances (*Fulham FC Ltd v Cobra Estates Plc* (1994)). There appears to be a dissonance between s 173(2) and that case law.

The duty to exercise reasonable care, skill and diligence

The statutory principle

There is a very difficult question as to the overarching standard of care that the director owes to the company at common law. The standard of behaviour and professionalism expected of company directors has changed through the 20th century in the common law as our perception of directors as full-time managers of the company has developed. Section 174 of the CA 2006 has now created a statutory duty to exercise reasonable care, skill and diligence in the following terms:

A director of a company must exercise reasonable care, skill and diligence.

That this is an obligation is indicated by the word 'must'. There is a question as to whether or not there is a need to exercise reasonable care or alternatively skill or alternatively diligence, or whether there is one single duty that is a duty to exercise reasonable care and also skill and diligence in one composite duty. The statute does not make this clear. The cases have not differentiated between these three terms and therefore it is supposed that there is one composite obligation imposed on directors.

An important question that has arisen in the case law is whether any given director must be measured subjectively against her own level of competence or whether that director should be measured against an objective concept of what constitutes reasonable care, skill and diligence for any director. Section 174(2) resolves this question in the following way:

> This means the care, skill and diligence that would be exercised by a reasonably diligent person with –
>
> (a) the general knowledge, skill and experience that may reasonably be expected of a person carrying out the functions carried out by the director in relation to the company, and
> (b) the general knowledge, skill and experience that the director has.

Therefore, what is measured is the objective knowledge of anyone acting as a director and also the knowledge, skill and experience of that individual director. A director with little experience will be measured against a standard of general knowledge of an ordinary director, even though that particular director does not have that skill or experience. It may also be that a given director may be an expert in mining but not understand financial matters: on appointment to the board of directors of a mining company, we might think it unfair to expect that director to be liable for errors in financial policy when that person's expertise was limited to mining. Nevertheless, all directors are required to live up to standards expected of an objective director. The requirement of 'reasonableness' in s 174(2)(a) would be the principal avenue for defending that director on the basis, perhaps, that the skill in question was beyond the reasonable skill-set of an ordinary director. Bound up with this duty is a need for individual directors to enhance their own skill-sets and experience to cope with their obligations. After all, if directors were able to escape liability on the basis that they did not have a particular skill or any knowledge, then that would encourage directors to remain ignorant and not involve themselves in the affairs of the company, which would be to the disadvantage of companies generally.

The principle in the case law

The roots of the principle in Re City Equitable Fire Insurance Co. Ltd

The case law dealing with this principle is best understood as having begun in its modern form with the decision of Romer J in *Re City Equitable Fire Insurance Co.*

Ltd (1925) and the retreat from that decision in subsequent cases. Romer J took the approach that directors bore little personal responsibility to turn up for anything more than board meetings and that they could delegate the day-to-day administration of the company's business to subordinate managers and employees. This seems like an odd approach to the duties of executive directors in the 21st century, because directors in the 1920s were effectively being excused from the need to do any work and from being liable for anything that resulted from their indolence.

In *Re City Equitable Fire Insurance Co. Ltd*, Bevan, who effectively controlled all of the other directors, committed widespread fraud and embezzlement from the company. This scandal was notorious in the 1920s. The legal issue was whether or not Bevan's fellow directors (and the company's auditors) should have been liable to the company for the losses that the company suffered as a result of those directors allowing Bevan to do what he did. Among the acts that were complained about were directors who signed blank cheques for Bevan and the auditors who allegedly overlooked fraud when scrutinising the company's books. It was held that the directors who had signed blank cheques in favour of third parties had not done so in circumstances in which there was anything to have caused them to be suspicious and so they were found not to be liable, which suggested that it was enough for directors to rest on their laurels and not be scrupulous in the performance of their duties.

Significantly for the general law on directors' duties, Romer J held 'a director must, of course, act honestly; but he must also exercise some degree of both skill and diligence' (at p 427). More specifically, his Lordship set out three propositions (which have been doubted in more recent cases and by s 174 of the CA 2006) as to directors' duties: first, that the level of skill was subjective; second, that the director is not required to give continuous attention to her duties; third, that a director may delegate her responsibilities in general terms. Each of these propositions is considered in turn.

The subjective level of skill

In accordance with the law at the time, Romer J held that a 'director need not exhibit in the performance of his duties a greater degree of skill than may reasonably be expected from a person of his knowledge and experience' (at p 428). This suggests a subjective approach to the standard of performance. As Lindley MR had held in *Lagunas Nitrate Co. v Lagunas Syndicate* (1899), 435:

> If directors act within their powers, if they act with such care as is reasonably to be expected from them, having regard to their knowledge and experience, and if they act honestly for the benefit of the company they represent, they discharge both their equitable as well as their legal duty to the company.

This means that a director was to have been held to her own, subjective standard of ability, knowledge and experience, and not that of the reasonable man. Such a director was required, however, to take the same level of care in the performance of her duties as an ordinary man might be expected to take when acting on her own behalf. This approach was very similar to the law of trusts at that time (*Learoyd v Whiteley* (1887)).

More recently, however, this principle has been restated by Hoffmann LJ in two cases. In *Norman v Theodore Goddard* (1991), it was held that directors should be assessed by the standard of care of a reasonably diligent person who had the knowledge, skill and experience both of a person carrying out that director's functions objectively and of the defendant subjectively. Then in *Re D'Jan of London Ltd* (1994), a director was accused of negligence when he had signed an inaccurate fire insurance proposal form to insure the company's property, with the result that the company was uninsured when its premises caught fire. The director signed the form without having read it. Hoffmann LJ clearly formed a dim view of the defendant, having seen him in the witness box (because Hoffmann LJ was here acting at first instance), and thus considered that the director 'did not strike [his Lordship] as a man who would fill in his own forms' (p 562). Hoffmann LJ considered that the standard of care required of a director as follows:

> I do not say that a director must always read the whole of every document which he signs. If he signs an agreement running to 60 pages of turgid legal prose on the assurance of his solicitor that it accurately reflects the board's instructions, he may well be excused from reading it all himself. But this was an extremely simple document asking a few questions which Mr D'Jan was the best person to answer.

Therefore, the director was personally liable to compensate the company because this was a simple form that the director should have read himself. There may, however, be circumstances in which a director might be absolved from liability.

The quality of attention that a director must give to the affairs of the company

Romer J also held that:

> A director is not bound to give continuous attention to the affairs of his company. His duties are of an intermittent nature to be performed at periodical board meetings, and at meetings of any committee of the board upon which he happens to be placed. He is not, however, bound to attend all such meetings, though he ought to attend whenever, in the circumstances, he is reasonably able to do so [at p 428].

A more modern and appropriate approach, it is suggested, is that set out in *Dorchester Finance Co. Ltd v Stebbing* (1989), in which two non-executive directors were held to be negligent in not attending the board meetings of a subsidiary company, even though it was shown that it was not commercial practice to do so. These directors had also signed blank cheques for a director who rarely attended the company's offices, with the result that that director embezzled money from the company. Foster J ostensibly approved the test set out by Romer J, although his Lordship applied that test in a different way. Foster J held that it was important to recognise that the directors in that case were accountants and people with long experience of acting in the financial affairs of various bodies, and consequently that they could not assert that they had no duties to perform as non-executive directors. Foster J was 'alarmed' at the very suggestion. The signing of blank cheques was also held to have been negligent. As for the director who misused the blank cheques, such that he took £400,000 from the company, it was held that he had 'knowingly and recklessly misapplied the assets' of the company with the effect that he had committed gross negligence and so was liable for 'damages'.

The extent to which a director may delegate responsibilities to another person

It was also held by Romer J that a director is entitled to trust a manager or other official to perform any duties which are delegated, provided that that is done honestly, that it complies with the articles of association, that it is appropriate for the company's business and that the director had no reason to suspect the delegate of committing a wrong if tasks were delegated to her. However, in *Equitable Life Assurance Society v Bowley* (2004), it has been suggested that this third proposition in *City Equitable Fire* no longer represents the modern law. Jonathan Parker J conceived of this principle in the following way in *Re Barings plc (No. 5)* (1999) at 489:

> (i) Directors have, both collectively and individually, a continuing duty to acquire and maintain a sufficient knowledge and understanding of the company's business to enable them properly to discharge their duties as directors. (ii) Whilst directors are entitled (subject to the articles of association of the company) to delegate particular functions to those below them in the management chain, and to trust their competence and integrity to a reasonable extent, the exercise of the power of delegation does not absolve a director from the duty to supervise the discharge of the delegated functions. (iii) No rule of universal application can be formulated as to the duty referred to in (ii) above. The extent of the duty, and the question whether it has been discharged, must depend on the facts of each particular case, including the director's role in the management of the company.

This approach seems to chime in more successfully with the modern operation of companies, particularly in large public companies where a greater number of activities need to be performed and therefore they may need to be delegated to more junior employees. In the 21st century we expect professional executive directors to take responsibility for the success or failure of a business. The *Barings* litigation dealt with senior management in a Barings Bank who failed to supervise the activities of employees in the Singapore branch, who were able to run up losses large enough to bankrupt the entire institution. In the modern world it is impossible to absolve those directors of responsibility for failing to put suitable processes in place to control their employees. When the case law exhibits an occasional tendency to excuse directors for liability in this sort of area, it should be remembered that regulation of financial institutions, for example, obliges a bank to have systems in place precisely so as to control the activities of all personnel who can affect the business of the institution (see the FSA Handbook's 'SYSC' rulebook).

The duty to avoid conflicts of interest

Introduction

One of the core tenets of the law on fiduciary duties is the obligation to avoid conflicts of interest. This is a core principle of equity (see Hudson 2009a, 12.5). The conflict of interest that is at issue is a conflict between a director's fiduciary duties and her personal interests.

If a fiduciary permits a conflict of interest, then that will have two consequences in equity. First, the fiduciary will be required to account for any unauthorised profit acquired as a result of a conflict of interest (*Boardman v Phipps* (1967)). This account takes the form of holding any profit on constructive trust for the beneficiaries of that fiduciary duty, or of holding any property acquired with that profit on constructive trust for the beneficiaries of that fiduciary duty, or (if the property or its traceable proceeds cannot be identified) of accounting personally to the beneficiaries of that fiduciary duty for the amount of the profit (*CMS Dolphin Ltd v Simonet* (2001)). Second, the fiduciary must not only avoid actually taking profits from such a conflict but must also prevent any possibility that there has even been a conflict of interest (*Bray v Ford* (1896), 51, per Lord Herschell). In consequence the general equitable principles have been applied very strictly indeed, as Lord Chancellor King intended in the early case of *Keech v Sandford* (1726).

Some recent decisions of the Court of Appeal have suggested that these fiduciary obligations might be capable of some dilution, as will be considered below. However, these recent decisions do not show sufficient appreciation for the importance of the traditional fiduciary principles. More recent decisions still suggest that the old principles will still be observed. We shall begin, however, with the principle as set out in the statute.

The statutory principle

The core duty

The duty to avoid conflicts of interest is expressed in s 175(1) CA 2006 in the following terms:

A director of a company must avoid a situation in which he has, or can have, a direct or indirect interest that conflicts, or possibly may conflict, with the interests of the company.

There is therefore a distinct obligation on each individual director. It also applies to former directors if that director became aware of the possibilities for exploiting any information or opportunity or property while a director (s 170(2)). The duty is not only to avoid taking a benefit from a conflict of interest, but it is broader than that. It is a duty to avoid situations where there may possibly be direct or indirect conflicts of interest. Therefore, the duty is breached prima facie simply by failing to avoid a situation where there may be a conflict of interest. It is important that it is the interests of the company that are at issue and not, for example, the personal interests of some of the shareholders or of any particular shareholder. (If the conflict of interest relates to a transaction with the company, that is governed by s 177 CA 2006 (s 175(3).)

The duty is defined as encompassing (but is not limited to) the situation set out in s 175(2):

This applies in particular to the exploitation of any property, information or opportunity (and it is immaterial whether the company could take advantage of the property, information or opportunity).

Therefore, an 'exploitation' of property, information or an opportunity – for example for financial gain – would constitute a conflict of interest, especially if the property or information was owned by the company, or if the opportunity was a maturing business opportunity which the company could have exploited. Importantly, the general principle in s 175(1) can apply beyond this sort of exploitation.

Avoiding liability where there has been authorisation

In the case law[,] the principal means of eluding liability for a conflict of interest has been to demonstrate that the fiduciary has received authorisation for acting on that conflict of interest. Section 175 contains an exception to the liability in s 175(1) based on acquiring authorisation in the manner set out in that section. Section 175(4) provides that:

This duty [in s 175(1)] is not infringed –

(a) if the situation cannot reasonably be regarded as likely to give rise to a conflict of interest; or
(b) if the matter has been authorised by the directors.

There are effectively two defences to a claim based on a conflict of interest. Thus paragraph (a) allows common sense to decide whether or not there is a conflict of interest. Alternatively, paragraph (b) provides for a defence where the directors have 'authorised' the particular matter at issue. In the case law, what constituted authorisation was a key part of the jurisprudence. Section 175 has streamlined the process for acquiring authorisation for company directors. It has been argued that what s 175 has achieved is the dilution of the fiduciary principles by making it easier for directors to acquire authorisation by following the statutory procedure.

The definition of what constitutes 'authorisation' is considered in s 175(5) in the following terms:

> Authorisation may be given by the directors –
>
> (a) where the company is a private company and nothing in the company's constitution invalidates such authorisation, by the matter being proposed to and authorised by the directors; or
> (b) where the company is a public company and its constitution includes provision enabling the directors to authorise the matter, by the matter being proposed to and authorised by them in accordance with the constitution.

The concept of authorisation therefore applies differently to private companies and public companies. In relation to a private company, the articles of association may prevent an authorisation being granted by the directors. If there is no such prevention of authorisation being granted in the private company's constitution, the directors may authorise something that might otherwise be a conflict of interest. By contrast, in relation to public companies the company's constitution must contain a power for directors to authorise what would otherwise be a conflict of interest.

It is not clear whether *all* the directors or merely a majority of the directors are required to provide authorisation: s 175(4) makes a reference to 'the directors' without specifying. The statute appears to anticipate simply that a majority of the directors are permitted to authorise a conflict of interest. Section 175(6) CA 2006 qualifies the giving of authorisation in the following terms, which appear to suggest that any quorum of directors permitted in the company's constitution is sufficient:

> The authorisation is effective only if –
>
> (a) any requirement as to the quorum at the meeting at which the matter is considered is met without counting the director in question or any other interested director, and
> (b) the matter was agreed to without their voting or would have been agreed to if their votes had not been counted.

Thus the statute provides a mechanism for directors avoiding liability stemming from conflicts of interest. In essence, s 175 provides that a director bears an obligation to avoid even potential conflicts of interest, although that duty does not exist if the directors have authorised the conflict of interest, or if there is not likely to be any reasonable conflict of interest. The case law, as discussed next, would also prevent the sort of abuse where the directors may seek to authorise some of their number to take secret profits from the company contrary to the interests of the shareholders and the company itself. One clear concern about the authorisation procedure under s 175 is the possibility that directors will approve a transaction entered into by another dominant director or some other person not because they genuinely believe it to be appropriate, but because they always obey that other person's requests. Such a clandestine authorisation might be difficult to prove in practice if the directors purported to grant their authorisation in formal compliance with s 175. It is suggested that the case law will remain important in situations like this.

The principle in the case law

The principle in essence

The House of Lords in *Keech v Sandford* (1726) is one of the earliest cases in which the principle of equity emerges that a fiduciary may not take unauthorised profits from that fiduciary office. The basis for this principle was held by the House of Lords in *Boardman v Phipps* (1967) to be that the fiduciary may not permit conflicts between his personal capacity and his fiduciary capacity. The correct approach to the nature of the company's rights when a director acquires an unauthorised profit was set out by Rimer J in *Sinclair Investment Holdings SA v Versailles Trade Finance Ltd (No 3)* (2007) and by Lawrence Collins J in *CMS Dolphin Ltd v Simonet* (2001), following the earlier House of Lords authorities, to the following effect. The company's primary right is to have the fiduciary's unauthorised profits, and any property acquired with those profits, held on constructive trust for the company. If there is no property that can be separately identified as being held on such a constructive trust, the director owes a personal obligation to the company to account in money or money's worth for the amount of the profits.

Rimer J held in *Sinclair Investment Holdings SA v Versailles Trade Finance Ltd (No 3)* in the following terms:

> any identifiable assets acquired by fiduciaries in breach of their fiduciary duty are, and can be declared to be, held upon constructive trust for the principal [that is, the company] (*Boardman v Phipps* (1967), *AG Hong Kong v Reid* (1994), *Daraydan Holdings Ltd v Solland* (2005)) . . . There will in practice often be no identifiable property which can be declared by the court to be held upon such a constructive trust, in which case no declaration will be made and the principal may at most be entitled to a personal remedy in the nature of an account of profits. In Boardman's case the court made a declaration that the shares that had been acquired by the fiduciaries were held on constructive trust (a proprietary remedy), and directed an account of the profits that had come into their hands from those shares (a personal remedy). Boardman's case can be said to have been a hard case as regards the fiduciaries, whose integrity and honesty was not in doubt; and it well illustrates the rigours of

the applicable equitable principle. The recovery by the trust of the shares was obviously a valuable benefit to it; and equity's softer side was reflected in the making of an allowance to the fiduciaries for their work and skill in obtaining the shares and profits. On the very different facts of *Reid's* case, there was no question of any such allowance being made.

Therefore, the position is clear: the company's primary right is for a proprietary constructive trust over the director's profits; the secondary right (if there is no property over which the constructive trust can take effect) entitles the company to a personal remedy in the form of an account of profits from the director; and third, the court may make some equitable accounting to reduce the amount of any such account if the court considers the circumstances to be appropriate for the equitable relief of such a defendant (*Markel International Insurance Co Ltd v Surety Guarantee Consultants Ltd* (2008)).

The proprietary constructive trust has the effect that the profits are held on constructive trust, and therefore any property acquired with those profits is also held on the terms of that constructive trust. This constructive trust right also means that even if those profits are mixed with other money, the beneficiary acquires a right to trace into that mixture or into any substitute property and to impose a proprietary right over it (*Westdeutsche Landesbank v Islington* (1996)). It is only if there is no property that can be subjected to the constructive trust or if that property has decreased in value that the defendant would be liable to account personally for the amount of the profits.

The strictness of the principle traditionally

The leading case of *Boardman v Phipps* (1967) illustrates the strictness of the principle that flowed originally from *Keech v Sandford*. In *Boardman v Phipps*, the beneficiaries suffered no loss when Boardman, a solicitor who advised their trustees, acquired shares in a private company on his own account while advising the trustees as to the trust's shareholding in that same company. Nevertheless, the beneficiaries were held to be entitled to force Boardman to account for the unauthorised profits that he made through his fiduciary office on those shares. The basis of his liability was predicated on the potential conflict between his personal interests and his fiduciary office.

The roots of the principle have recently been explained as being in equity and in the need to prevent fiduciaries from acting unconscionably by permitting conflicts of interest (*Yugraneft v Abramovich* (2008), 373, per Clarke J). As this principle was expressed by Lord Herschell in *Bray v Ford* (1896), 51:

> It is an inflexible rule of the court of equity that a person in a fiduciary posi-
> tion . . . is not, unless otherwise expressly provided, entitled to make a profit;
> he is not allowed to put himself in a position where his interest and duty
> conflict. It does not appear to me that this rule is, as had been said, founded

upon principles of morality. I regard it rather as based on the consideration that, human nature being what it is, there is danger, in such circumstances, of the person holding a fiduciary position being swayed by interest rather than by duty, and thus prejudicing those whom he was bound to protect. It has, therefore, been deemed expedient to lay down this positive rule.

So, the rule is considered to be a strict rule, as was accepted by separately constituted Houses of Lords in *Aberdeen Railway Co v Blaikie Bros* (1854) and *Regal v Gulliver* (1942) in relation specifically to company directors; and in *Boardman v Phipps* (1967). However, it is worthy of note that in a little-quoted passage, Lord Herschell did go on to say that (p 52):

> I am satisfied that it might be departed from in many cases, without any breach of morality . . . Indeed it is obvious that it might sometimes be to the advantage of the beneficiaries that the trustee should act for them professionally rather than a stranger, even though the trustee were paid for his services.

However, it is the strict approach which has been relied upon in leading cases such as *Regal v Gulliver* and *Boardman v Phipps*, which are considered next. Indeed, it may be that Lord Herschell only intended an exception to operate where the trustee is properly authorised to act in such a capacity or where the beneficiaries agree that she should act in this way, in which case there would be authorisation and so no liability in any event.

The application of this principle in Regal v Gulliver *and in* Boardman v Phipps

The House of Lords in *Regal v Gulliver* (1942) considered a situation in which four directors of a company, which operated a cinema, sought to divert a business opportunity to acquire two further cinemas to a separate company which they controlled. The first company could not afford to acquire the rights to all of the cinemas out of its own resources and its four directors were reluctant to give personal guarantees for the extra sums needed. Therefore, those four directors (and a solicitor) subscribed for shares in the second company, which consequently took up the opportunity to operate the two further cinemas. So, in essence, the directors of a company had decided to divert an opportunity away from the company for which they worked into the second company so that they could keep the profits for themselves.

It was held that the directors' profits had been made from their fiduciary offices as directors. Therefore, they were required to account for those profits to the company. Lord Russell was clear that the obligation to account for profits was in no way predicated on proof of fraud or mala fides on the part of the directors. Instead, Lord Russell held that '[t]he liability arises from the mere fact of a profit having, in the stated circumstances, been made' by a fiduciary.

The four directors were not able to rely on their purported grant of authorisation to themselves to pursue this opportunity on their own account. As considered above, under s 175 CA 2006 it is now possible to acquire authorisation from the directors following the procedure set out in the statute. However, it was held in *IDC v Cooley* (1972) and in *Gwembe Valley Development Co Ltd v Koshy (No 3)* (2004) that if a director fails to make a sufficiently full disclosure of all the relevant facts when seeking authorisation from the other directors, that director will not be taken to have acquired authorisation to earn personal profits.

The proof of authorisation in the case law

There is, effectively, a defence to an action for constructive trust on grounds of making personal profits that the fiduciary had authorisation so to do. The effect of s 175(5) of the CA 2006 is that it formalises the way in which fiduciaries can acquire the necessary authorisation, whereas on the authorities there were few circumstances in which authorisation was found to have been granted. It was suggested by the House of Lords in *Boardman v Phipps* that a solicitor advising a trust could have avoided liability had he made a disclosure of his intention to make share purchases on his own account, whereas on the facts no such disclosure was made. Similarly, in *Regal v Gulliver* there was no disclosure and therefore no authorisation, except in the form of a purported authorisation by the directors to themselves for the action that they proposed to take. It was held, in essence, that the directors could not authorise themselves. The CA 2006 has changed this position for company directors because directors may now acquire authorisation from the other directors in accordance with s 175(5). The cases on authorisation may nevertheless retain some significance, particularly in circumstances in which it might be contended that the director failed to make an appropriate disclosure to the other directors or otherwise acted unacceptably.

For example, in the Privy Council decision in *Queensland Mines v Hudson* (1977), the defendant had been managing director of the plaintiff mining company and had therefore been in a fiduciary relationship with that company. The defendant had learned of some potentially profitable mining contracts. The board of directors of the company decided not to pursue these opportunities after they had had all of the relevant facts explained to them. The board of directors decided not to pursue the opportunity in the knowledge that the managing director intended to do so on his own account. Importantly, what happened was that one director had received authorisation with the informed consent of the remainder of the board. Therefore, when the managing director resigned and made profits from that opportunity, the company sought to recover the profits that he then earned. It was held that the managing director had obtained authorisation.

By contrast, in *Industrial Development Consultants Ltd v Cooley* (1972), a managing director was offered a deal by one of the company's clients provided that he worked for them on his own account and not through the company. Without disclosing this fact to the company and claiming to be in ill health, the managing director resigned and entered into a contract with the third party within a week of his resignation. It was held that the managing director had occupied a fiduciary position in relation to his employer company throughout. He was therefore required to disclose all information to the company and to account for the profits he made under the contract. It was significant, in the judgment of Roskill J, that the director had misled his employer about his reasons for leaving on these facts, having pretended to be resigning on account of his ill health.

Similarly, in *Crown Dilmun v Sutton* (2004) a director learned of an opportunity to develop a football ground, which he exploited on his own account once his contract of employment had been terminated. It was found that he had not made full disclosure to the claimant company of the extent of the opportunity. Consequently, he was liable to account to the claimant for the personal profits realised from the transaction. In *Item Software (UK) Ltd v Fassihi* (2003) it was held that if a director sought to tempt a client away from her employers, then that would constitute a breach of fiduciary duty with the effect that any profits so earned would have to be accounted for by way of constructive trust to the employer company.

What remains unclear is whether the case law principles will continue to apply over and above the means of acquiring authorisation under s 175, for example relying on the *dicta* of Lord Russell in *Regal v Gulliver* to the effect that a vote by the shareholders in a general meeting would constitute authorisation (even if, for example, the directors had voted against the transaction, or if all of the directors stood to take a benefit from the transaction and so there were no directors without a conflict of interest who could vote). It is suggested that this general equitable principle, that the company (by means of a shareholder vote) can ratify the directors' actions, ought to continue to apply in such circumstances even after the passage of the CA 2006.

Eluding conflicts of interest: the corporate opportunity doctrine

The case law in company law had sought to drift away from mainstream fiduciary law even before the enactment of the CA 2006. This section considers the way in which a line of company law cases began to divert from mainstream fiduciary law; and the next section considers how more recent cases have reverted to the mainstream principles. In essence, a line of company law cases began to suggest that if the director were not diverting a business opportunity away from the company, there would not be a conflict of interest. This change in the company law cases was on the basis that company directors operate in a commercial context that requires them to be able to act differently from other forms of fiduciary. This concept may lead to an argument that much directorial activity is outside the scope of s 175 CA 2006.

So, in *Island Export Finance Ltd v Umunna* (1986), it was held, in effect, that the director did not divert a business opportunity from the company and therefore that the conflict of interest principle did not apply. In that case, the company had a contract with the government of Cameroon to supply postboxes. Mr Umunna resigned from the company once the contract was completed, having worked on that contract and acquired a great deal of expertise in that particular activity. The company ceased pursuing this line of business and, after his resignation, Mr Umunna entered into a similar contract on his own behalf. The company sued him for the personal profits that he made for himself under this second contract. It was held that Mr Umunna did not divert a business opportunity away from the company because the company had ceased working in that field and Mr Umunna was relying entirely on his own contacts and experience.

What was also held in *Island Export Finance* was that a fiduciary's obligations to the company continue even after the director has resigned. This makes sense, or else it would be all too easy to avoid your obligations simply by resigning from a company. Nevertheless, that does not prevent a director from beginning new activities after leaving the company. So, in *Balston v Headline Filters Ltd* (1990), a director had resigned from a company and leased premises with a view to starting up in business on his own account before a client of the company approached him and asked him to work for them. Falconer J held that there was no breach of duty in these circumstances because there was nothing wrong with a director leaving the company and setting up in business on his own account. Moreover, there had not been any maturing business opportunity in this case that the director had diverted to himself.

Perhaps the most troubling development in the case law was the attitude of the courts to situations in which directors had been marginalised within the company before deciding to exploit an opportunity on their own account. So, in the case of *In Plus Group Ltd v Pyke* (2002), Mr Pyke was a director of a company, In Plus Ltd, but he had 'fallen out with his co-director' and in consequence he had been 'effectively excluded from the management of the company'. Mr Pyke decided to set up a company on his own while he was still a director of In Plus Ltd, and that company entered into contracts on its own behalf with a major customer of In Plus Ltd. Remarkably, the Court of Appeal held that Mr Pyke was not in breach of his fiduciary duties to In Plus Ltd because he had not used any property belonging to In Plus Ltd and also because he had not made any use of any confidential information which he had acquired while he was a director of In Plus Ltd. As Sedley LJ held:

> Quite exceptionally, the defendant's duty to the claimants had been reduced to vanishing point by the acts (explicable and even justifiable though they may have been) of his sole fellow director and fellow shareholder Mr Plank. Accepting as I do that the claimants' relationship with Constructive was consistent with successful poaching on Mr Pyke's part, the critical fact is that it was done in a situation in which the dual role which is the necessary predicate of [the claimants'] case is absent. The defendant's role as a director of the claimants was throughout the relevant period entirely nominal, not in the sense in which a non-executive director's position might (probably wrongly) be called nominal but in the concrete sense that he was entirely excluded from all decision-making and all participation in the claimant company's affairs. For all the influence he had, he might as well have resigned.

Therefore, while this is expressed as being an exceptional decision, it seems that if you have been marginalised within the company, then you are entitled to compete with it from your desk within that company. This is exactly the contrary of what was intended in leading cases such as *Bray v Ford* and *Regal v Gulliver*.

Other cases have also suggested a dilution of the principle. So, in *Murad v Al-Saraj* (2005), the claimants entered into a joint venture with the defendant to buy a hotel. The defendant was found to owe fiduciary duties to the claimants. However, the defendant did not disclose the personal profits that he stood to earn from the transaction. While the Court of Appeal held that those profits were to be held for the claimants, Arden LJ chose to cast doubt on the suitability of the doctrine in *Boardman v Phipps* on the basis that it imposed liability on people who had not been demonstrated to have acted wrongly. She expressed a preference for liability being based on some fault by the defendant. This completely misunderstands the purpose of the conflict of interest principle. First, it is a fault for a fiduciary to permit a conflict of interest, as indicated in *Bray v Ford*. Second, the purpose of the doctrine is to prevent conflicts of interest and not to allocate liability for compensation for fault. Third[,] and notably, s 175 CA 2006 does not take the approach used by Arden LJ.

Nevertheless, Rix LJ took a similar approach in *Foster v Bryant* (2007) to that of Arden LJ. His Lordship, with respect, became overly concerned in his survey of the cases with the idea that the defendant must be misusing company property before being held liable to hold unauthorised profits on constructive trust. This overlooks the central point made by Lord Upjohn in *Boardman v Phipps* (1967) that the rule against unauthorised profits was intended to prevent both actual conflicts of interest and also the possibility that there might be a conflict of interest, as enshrined in s 175 CA 2006.

In *Foster v Bryant* the defendant director was found to have been excluded from the operation of the business, just as in *In Plus*. While still a director, the defendant entered into negotiations with one of the company's principal clients to work directly for them. The company contended that the defendant had been a director of the company when the business opportunity came to light and therefore that any profits earned from that opportunity should be held for the company. Rix LJ held that the earlier cases that had imposed this sort of liability had done so in relation to 'faithless fiduciaries'. (This is not the case: the principle is based on avoiding conflicts of interest.) Nevertheless, Rix LJ held that the defendant had not behaved inappropriately on these facts.

It is suggested that the developments mooted in these cases are unfortunate and should be resisted.

Reaffirming the traditional principle

It has been held in company law that company directors may leave their posts and begin trading on their own account. The real issue is how much time must pass and in what circumstances a director may begin trading on her own account.

Significantly, however, a director may not, even after resigning from her post, use either the company's property or information which she had acquired while still a director of the company to generate personal profits (*Ultraframe UK Ltd v Fielding* (2005)). If this proviso were not imposed, then it would be very easy to elude the duty to avoid conflicts of interest simply by learning of an opportunity while a director with the company, resigning immediately, and then exploiting that opportunity the following day on your own account.

In *CMS Dolphin Ltd v Simonet* (2001), a director of a company fell out with his co-director and began to divert the company's clients to a new business he set up on his own account. Lawrence Collins J upheld Simonet's liability for diverting a corporate opportunity and so made him liable to account as a constructive trustee for the profits that had been earned from this activity. What is particularly important for present purposes is the manner in which Lawrence Collins J concluded his judgment by explaining the operation of this doctrine, as his Lordship saw it:

> In my judgment the underlying basis of the liability of a director who exploits after his resignation a maturing business opportunity of the company is that the opportunity is to be treated as if it were property of the company in relation to which the director had fiduciary duties. By seeking to exploit the opportunity after resignation he is appropriating for himself that property. He is just as accountable as a trustee who retires without properly accounting for trust property. In the case of the director he becomes a constructive trustee of the fruits of his abuse of the company's property, which he has acquired in circumstances where he knowingly had a conflict of interest, and exploited it by resigning from the company.

Therefore, in summary, if a director diverts a maturing business opportunity away from the company to himself personally or to some entity under his control, then he will be required to account for any profits made by way of constructive trust and to account to the company for any loss suffered by the company as a result. It is suggested that this approach must be correct in principle.

The question is whether the pure principle of avoiding conflicts of interest was being recognised in the cases, or whether the idea of conflicts of interest is to be interpreted on the narrow basis set out in *Foster v Bryant*. It is suggested that *Foster v Bryant* and *Murad v Al-Saraj* should be understood as being aberrations. Three more recent cases have reverted to the traditional approach.

The Court of Appeal in *Re Allied Business and Financial Consultants Ltd* (2009) returned to the traditional 'no conflict' principle in *Aberdeen Railway* and *Regal v Gulliver* to the effect that company directors owe an undivided loyalty to the company akin to the obligations of a trustee. Here a director had diverted a business opportunity to an undertaking under his control when it had been difficult to interest any of the company's clients in investing in the opportunity. The strictness of the traditional principle was nevertheless upheld.

A comparatively straightforward case of diverting business opportunities away from a company arose in *Berryland Books Ltd v BK Books Ltd* (2009). However, this is not simply a case involving diversion of a business opportunity, but rather it also involved the use of the company's assets. The defendant director managed the UK operation of an international publishing company. The director then set up a new company that took over the first company's business opportunities and solicited its staff to work for the new company. In essence, the defendant used all of the first company's assets and staff under a new company name, even to the extent of using the first company's resources to set up competing websites and to market the new company's products at industry conferences. It was held that the defendant had breached the 'fundamental duty of loyalty' bound up with being a director.

In *PNC Telecom plc v Thomas (No2)* (2008) it was held that there would be a conflict of interest when the defendant director voted in a situation where there was a clear conflict of interest: the vote related to a proposal for a change in control of the company which would affect whether or not that director was removed as a director. It was held that the conflict of interest principle still applied, even though this was clearly the sort of case on exclusion from participation in the management of the business that had motivated Rix LJ in *Foster v Bryant*.

In reading each of these cases, it is clear that the judges could have adopted the trend established by Sedley, Arden and Rix LJJ in the cases discussed above. In each case, it could possibly have been argued that the directors had been excluded from the management of the business or that they had sought to leave the directorship before exploiting an opportunity. Alternatively, Rix LJ might even have considered that the defendant in *Berryland Books* was a 'false fiduciary'. However, the important point is that in each case the courts returned to the leading cases and to the strict version of the rule.

Should directors' fiduciary duties be developed in a commercial context?

Traditionally the case law before 1897 had conceived of fiduciary duties as imposing one-dimensional obligations of the utmost good faith on directors, trustees and others. However, the suggestion emerged in the late 20th century that the many commercial factors that made demands on directors might mean that those duties should be understood more contextually, and that directors' duties should consequently be seen in a more commercial light. In this vein, many commentators have argued that directors should be understood as being dynamic commercial managers rather than as being ordinary trustees, and therefore it is said that they should bear different duties from those owed traditionally by trustees.

In my opinion this would be an undesirable development. For the reasons discussed in Chapters 14 and 15 at the end of this book, I suggest that the commercial world now occupies a different position from the one that many people thought it occupied in the 1980s. There is now a moral context to large commercial undertakings brought on by concerns about the environment, about the exploitation of the developing world, and about the rights of employees in companies. Consequently, it goes against the grain to argue that company directors ought not to be bound by the traditional fiduciary duties of the utmost good faith towards the company and its stakeholders. When we recognise that the commercial and the natural worlds are changing rapidly, the appropriate response is to require commercial people to have regard to the values which should guide their actions and decisions in that changing world, not to dilute those values so that they can lurch from one reckless decision to another.

Traditional fiduciary duties are a key part of the values that directors should observe as fiduciaries. A director should not argue that she wants to be allowed to do whatever she wishes. Instead, the director should recognise that she has always been bound by obligations to act selflessly in the interests of the company, to avoid conflicts of interest.

The duty not to accept benefits from third parties

The statutory principle

It is a part of equity that a fiduciary may not accept a bribe. Section 176 CA 2006 provides that a director must not accept benefits from third parties in the following terms:

(1) A director of a company must not accept a benefit from a third party conferred by reason of –
(a) his being a director, or
(b) his doing (or not doing) anything as director.

In this context, s 176(2) provides that the term 'third party' means 'a person other than the company, an associated body corporate or a person acting on behalf of the company or an associated body corporate'. However, a director does not take a benefit if that benefit is paid 'from a person by whom his services (as a director or otherwise) are provided to the company are not regarded as conferred by a third party' (s 176(3)). The kernel of the directors' liability is based on conflicts of interest under s 176. Consequently, s 176(4) of the CA 2006 provides that '[t]his duty is not infringed if the acceptance of the benefit cannot reasonably be regarded as likely to give rise to a conflict of interest'. Section 176 applies to former directors as well as to current directors (s 170(2)).

The principle as developed in the case law in relation to bribes

As considered above, the principle in s 176 is based on the case law that has come before it. The Privy Council in *Attorney-General for Hong Kong v Reid* (1994) established the modern principle that a bribe received by a fiduciary must be held on constructive trust from the moment of its receipt and consequently that any property acquired with that bribe is also held on constructive trust. Lord Templeman held that a proprietary constructive trust is imposed as soon as the bribe is accepted by its recipient, with the effect that the company is entitled in equity to any profit generated by the cash bribe received from the moment of its receipt. Thus, the bribe and anything acquired with the bribe is held on constructive trust. Furthermore, Lord Templeman held that the constructive trustee is liable to account to the beneficiary for any *decrease* in value of the investments acquired with the bribe. Lord Templeman considered bribery to be an 'evil practice which threatens the foundations of any civilised society'. As such, the imposition of a proprietary constructive trust was the only way in which the wrongdoer could be fully deprived of the fruits of their wrongdoing.

It has been held latterly in *Yugraneft v Abramovich* (2008), 373 that the rationale for the constructive trust in *Reid* was that it would have been 'as a fiduciary unconscionable for him to retain the benefit of it', even though that was not the precise rationale used by Lord Templeman. Lord Templeman considered that the bribe should have been deemed to pass to the beneficiary of the fiduciary power at the instant when it was received by the fiduciary, and therefore that a proprietary right to the bribe passed automatically on constructive trust to the beneficiary of the arrangement (that is, the company if the fiduciary is a company director).

This principle was followed in *Tesco Stores v Pook* (2003), where the fiduciary had tendered invoices to the company for services that had not in fact been rendered. One of these amounts was a bribe that had been paid to the defendant. It was held by Peter Smith J that the bribe should be held on constructive trust. Similarly, in *Daraydan Holdings Ltd v Solland International Ltd* (2005), payments equivalent to bribes were paid as part of an inducement in a property development transaction. Lawrence Collins J held that the fiduciary in this transaction should not have put themselves in a situation in which their duty and their personal interests conflicted, and therefore that a constructive trust would be imposed over the bribes. Thus, the constructive trust here was based on avoidance of conflict of interest as well as the need to deal with the 'evil practice' of bribery.

The duty to declare an interest in a proposed transaction

The statutory principle

Aside from the general principles relating to the avoidance of conflicts of interest, the CA 2006 has developed specific principles in relation to situations in which a director has a personal interest in a proposed transaction with the company over which they are a director. In this regard, s 177 provides that:

(1) If a director of a company is in any way, directly or indirectly, interested in a proposed transaction or arrangement with the company, he must declare the nature and extent of that interest to the other directors.

What it means to have an interest in a transaction is not defined in the legislation, and so it is suggested that it should be interpreted broadly. The statute identifies one means by which the director may make this declaration. Section 177(2) CA 2006 provides that:

(2) The declaration may (but need not) be made –
    (a) at a meeting of the directors, or
    (b) by notice to the directors in accordance with –
        (i) section 184 (notice in writing), or
        (ii) section 185 (general notice).

Therefore, the director may either make a declaration at a meeting of the board of directors (as suggested by the plural 'the directors') or by means of a notice under s 184 or s 185 CA 2006.

It is important that the director involved must keep the other directors informed of any changes in circumstance. Section 177(3) provides that '[i]f a declaration of interest under this section proves to be, or becomes, inaccurate or incomplete, a further declaration must be made'. Significantly, all such declarations must be made *before* the company enters into the transaction (s 177(4)). The director does not face liability if the director was not aware of the nature of the transaction (s 177(5)) or if it does not reasonably involve a conflict of interest (s 177(6)).

The principle in the case law

The basis of the original equitable principle was expressed in the following *dicta* of Lord Cranworth in *Aberdeen Railway Co v Blaikie Bros* (1854):

it is a rule of universal application, that no one, having such [fiduciary] duties to discharge, shall be allowed to enter into engagements in which he has, or can have, a personal interest conflicting, or which possibly may conflict, with the interests of those whom he is bound to protect.

The effect of s 177 of the CA 2006 is that a director may be involved in such a transaction provided that a declaration of that interest has been made in the manner considered above before the transaction was created. Before the CA 2006 was enacted, the application of fiduciary principles to directors allowed a company automatically to avoid any contract which the board entered into on its behalf in which one or more of the directors had an interest, unless that interest had been disclosed to the company and approved by the general meeting (*Aberdeen Railway Co v Blaikie Bros* (1854)). This was so whether or not the director was acting bona fide for the benefit of the company. Any benefit derived by the director from such a contract could also be recovered (*Parker v McKenna* (1874)). This is, broadly speaking, the effect of s 177 as discussed above. Under the case, a transaction needed to be disclosed if the director had an interest, for example, as a shareholder (*Transvaal Lands Co. v New Belgium (Transvaal) Land, Co.* (1914)) or as a partner (*Costa Rica Ry v Forwood* (1901)). In *Cowan de Groot Properties Ltd v Eagle Trust* (1991), it was suggested that a contract of sale between two companies where the defendant was a director of one company and a creditor of the other company would not normally require a disclosure. Ordinarily, the burden of proving good faith in transacting and disclosing should be borne by the directors (*Newgate Stud Company v Penfold* (2004)).

There is a question as to whether or not disclosure can be implied in a given set of circumstances, for example as in *Lee Panavision v Lee Lighting* (1992) where all of the directors knew of the conflict of interest at issue. The Court of Appeal did not want to impose liability where the facts were known to all of the fiduciaries. However, this raises the problem that some of the shareholders may not know of the conflict of interest or that someone who becomes a member of the company later might not be able to find out about it if there was no formal declaration made. Lightman J took the view in *Re Neptune (Vehicle Washing Equipment) Ltd* (1995) that all such conflicts should be disclosed and recorded in the minutes, even if the director affected was the only person present at the meeting. This approach must be preferable.

The consequences of a breach of duty

Under the general law, as considered in relation to the various duties above, if the directors breach their duties then they will be liable to account for any profits made and to compensate the company for any loss incurred. If the director actually obtains corporate assets for herself, then she becomes a constructive trustee of those assets and the company will be able to recover the property or its proceeds from her.

It is provided in s 178(1) of the CA 2006 that the consequences of any breach or threatened breach of a director's general duties under ss 171 to 177 of the CA 2006 'are the same as would apply if the corresponding common law rule or equitable principle applied'. Except for the duty in s 174 to exercise reasonable care, skill and diligence, which takes effect as a common law principle, the directors' general duties are 'enforceable in the same way as any other fiduciary duty owed to a company by its directors' (s 178(2)).

Section 180 of the CA 2006 deals with consent, approval or authorisation by the company's members. In cases where the duty to avoid conflicts of interest or the duty to declare an interest in a proposed transaction or arrangement is complied with, then 'the transaction or arrangement is not liable to be set aside by virtue of any common law rule or equitable principle requiring the consent or approval of the members of the company' (s 180(1)).

Chapter 8

Shareholders' rights

Introduction

This chapter is concerned with the rights of shareholders. The greater part of this chapter is concerned with the way in which minority shareholders can protect themselves against the actions of the majority. This is at the heart of the law on shareholder rights because the basic proposition is that the majority shareholders can vote for whatever they want, and therefore the most interesting legislation and case law have related to the minority shareholders trying to resist a decision of the majority. Two further issues are also considered: first, the power to remove directors, and second, the legal basis on which dividends are paid to shareholders. Chapter 4 considered the rights of shareholders as they relate specifically to the company's constitution.

At the outset, we should remind ourselves of how a company functions in practical terms, and how company law interacts with that. In Chapter 6 we considered the operation of the company, and in Chapter 5 we considered the various forms of shareholding in the company. In those chapters we learned that different classes of shareholder have different types of rights and also that a large part of the company's business is conducted formally at the annual general meeting. Otherwise, the company is controlled on a day-to-day basis by the directors (and other senior management outside the board of directors if it is a large company). The operation of any trading company will involve making choices between, for example, investing for the long-term benefit of the business or generating short-term profits that can be distributed among the shareholders.

Therefore, there may be situations in which individual shareholders, or even groups of shareholders, feel that they have been badly treated either by the directors or by the votes of other shareholders. For example, the directors may wish to retain a year's profit in the business to invest in its future growth, instead of paying it out to the shareholders in dividends, whereas some of the shareholders may prefer to have that profit paid out so that they can receive dividends out of those profits in the short term. This will clearly generate conflict between directors and shareholders, or between different groups of shareholders. Different groups of shareholders may have different opinions as to the best future for the company and therefore one group of shareholders may vote at an annual general meeting for a course of action that is harmful to the interests of another group of shareholders. Again, this will generate conflict.

This chapter is primarily concerned with the resolution of these sorts of issues, in particular under s 260 and s 994 of the Companies Act 2006 (CA 2006), and s 122 of the Insolvency Act 1986. In later chapters, particularly in Chapter 10, we will consider the rights of shareholders in relation to the company's capital. In this chapter, however, we shall begin by considering the so-called rule in *Foss v Harbottle* (1843), which enshrines the principle of shareholder democracy: this means that the majority shareholders can vote for whatever they want at a company meeting, in accordance with the company's constitution; and they can vote to amend the constitution if the appropriate procedures are followed.

The rule in *Foss v Harbottle*: the principle of majority rule

It is a core principle of English company law that if a wrong is done to a company, it is the company itself that is the proper claimant and not the shareholders nor any group of shareholders. This follows from the idea that a company is a separate legal person with its own rights and responsibilities. Clearly, it may well seem to the shareholders in such a company that it is they who have suffered loss as opposed to the abstract entity that is the company. However, because companies are distinct legal persons in law, it follows that it must be the company itself that suffers a loss and that it is the company alone that is entitled to bring any action necessary to recover that loss. This may have the effect in certain circumstances that majority shareholders, who have in some way benefited from a transaction that otherwise may have been to the detriment of the company, may seek to ratify the transaction even if it has caused the company loss. Unsurprisingly, the minority shareholders who have not taken any of this benefit personally are likely to object to the transaction and its ratification by the majority of the shareholders. Nevertheless, the general principle remains that it is the company that is the proper claimant, and therefore if the majority of the shareholders vote to ratify the transaction, that will constitute the opinion of the company in law, subject to what is said below about oppression of the minority shareholders.

It is a second core principle of company law, closely linked to the first, that the majority of shareholders may do what they wish in relation to the company by exercising their voting power. The rationale behind this principle is that if shareholders representing the minority sought to resist or overturn anything for which the majority of shareholders had already voted, the majority shareholders would simply be able to vote again so as to ratify their earlier decision. Therefore, UK company law has taken the view that the majority viewpoint should always be enforced so as to prevent vexatious litigation by minority shareholders, except in specific circumstances considered in the next section.

This principle is commonly referred to as 'the rule in *Foss v Harbottle*'. In *Foss v Harbottle* in 1843 it was held that 'corporations like this . . . are in truth little more than private partnerships' and that issues within companies of this sort should be decided in accordance with the partnership agreement between the members. The model at issue here was clearly an old-fashioned joint stock company and not a post-*Salomon* company with separate legal personality. Nevertheless, it was held that even under this old model of company, corporate decision-making would depend on a vote under the constitution that bound all of the members together. This approach was applied by Lord Cottenham LC in *Mozley v Alston* (1847) to the effect that if injury is caused to the undertaking, it is caused to the corporation and not to its members in their private capacities. Therefore, even before the decision in *Salomon v A Salomon & Co Ltd* (1897), a concept of injury to the company was alive in the common law such that it was difficult for members of the company to demonstrate that they had suffered loss personally as a result of action of the members, even if that action had caused their shareholdings to fall in value. It was for the company to recover the loss and thus the members' indirect loss would be made good indirectly by the company's assets increasing by the amount of any compensation.

The principle which emerged from that rule was that the company should act on decisions which were made by the majority of the members: in plain English that means that a vote of the majority shareholders becomes the wish of the company itself, and the minority shareholders cannot ordinarily reverse it, nor complain about it afterwards (assuming that the vote was properly carried out and subject to the statutory protections discussed below). It was in *MacDougall v Gardiner* (1875) that Mellish LJ held the following, explaining why the minority should not be able to bring actions overturning the wishes of the majority:

> In my opinion, if this thing complained of is a thing which in substance the majority of the company are entitled to do, or if something has been done irregularly which the majority of the company are entitled to do regularly, or if something has been done illegally which the majority of the company are entitled to do legally, there can be no use in having litigation about it, the ultimate ends of which is only that a meeting is to be called, and then ultimately the majority gets its wishes. Is it not better that the rule should be adhered to that if it is a thing which the majority are the masters of, the majority in substance shall be entitled to have their will followed?

Consequently, we see the justification for this principle: it is based on the idea that it would be pointless to allow the minority shareholders to challenge a decision of the majority shareholders if the majority shareholders had done something which they would be legally entitled to ratify subsequently. However, his Lordship did also hold that:

> Of course if the majority are abusing their powers, and depriving the minority of their rights, that is an entirely different thing.

Importantly, then, minority shareholders may have a right of action if the majority are treating the minority shareholders in a way that is an abuse of their powers. In essence, this is the basis on which many of the cases suppressing actions of the majority have proceeded. Otherwise, however, the general principle remains, as set out by Lord Davey in *Burland v Earle* (1902), 'that the court will not interfere with the internal management of companies acting within their powers'.

The general rule derived from *Foss v Harbottle* (1843) is therefore clearly established and so the bulk of our attention will be on the circumstances in which that rule may be abrogated. The structure of our discussion shall be as follows. Having considered some general case law dealing with rights of shareholders, we shall consider three statutory mechanisms by which minority shareholders can object to the actions of the majority. First, we shall consider derivative actions under s 260 of the CA 2006. Second, we shall consider actions relating to unfair prejudice on the minority under s 994 of the CA 2006. Third, we shall consider the winding up of a company on the just and equitable grounds under s 122 of the Insolvency Act 1986. In each case we will begin by setting out the statutory provisions before considering the case law on that provision (or its predecessor, as appropriate). It is suggested that, for student readers, this may also be the most useful way of approaching problem questions in this area.

Derivative Actions

Bringing a derivative action: s 260 CA 2006

The first basis on which the will of the majority shareholders can be resisted is by means of a 'derivative action'. As was discussed above, it is the company that is the proper claimant when any wrong is done to the company or any abuse of any right is committed in relation to the company. However, if the majority shareholders are committing a wrong, it is likely that they will use their voting power to prevent the company from commencing litigation against them personally. This could mean, for example, that shareholders are induced to invest money in a company before the majority shareholders use their voting powers to starve those new shareholders of any dividends or to frustrate their reasonable ambitions for the company. Clearly, this principle could lead to abuse. Therefore, Parliament created a mechanism whereby even minority shareholders may take control of the reins and force the company to commence litigation in appropriate circumstances. This mechanism is contained in s 260 of the CA 2006 and is known as a 'derivative action'. This statutory mechanism took the place of the original rule in *Foss v Harbottle*. Since 2006, a derivative action may only be brought under the procedure in s 260 and not by any other means (s 260(2)). The basis for a derivative action is set out in s 260(3) in the following terms:

> A derivative claim under this chapter may be brought only in respect of a cause of action arising from an actual or proposed act or omission involving

negligence, default, breach of duty or breach of trust by a director of the company.

The cause of action may be against the director or another person (or both).

There are, therefore, two basic requirements. First, there must have been some default on the part of a director. Second, that default must have involved one of the four heads of claim. The four heads of claim are: negligence, default, breach of duty or breach of trust. It is suggested that 'negligence' should be taken to mean the ordinary tort of negligence at common law. It is also suggested that the ground of 'default' is potentially very broad indeed. It could encompass any default under the CA 2006 or under the common law. Similarly, a 'breach of duty' could encompass a breach of one of the directors' fiduciary duties or any other duty under the Act. A 'breach of trust' in relation to a director would seem to suggest specifically a breach of fiduciary duty. It is provided that the term 'director' in this context includes any former director and any shadow director of the company (s 260(5)).

Aside from the substantive issues as to the basis for bringing the derivative claim, there is also the procedural question as to the manner in which the claim must be brought. Section 261(1) provides that:

> A member ... who brings a derivative claim ... must apply to the court for permission ... to continue it.

Therefore, a derivative action requires the court's permission before it can even be commenced. Given the basis of the principle in preventing vexatious litigation in the first place, it is unsurprising that the court's permission is required before such an action can be brought. Section 261(2) further provides that the claimant must provide the courts with evidence to substantiate the claim that if that evidence does not disclose a prima facie case, the court must dismiss the application. Therefore, the applicant bears the burden of demonstrating on the evidence that there is at least some case to answer, on pain of having her application dismissed completely otherwise. Under s 261(3), if the claim is not to be dismissed, the court has the power to give directions as to the evidence that the company must provide and also has the power to adjourn proceedings to enable that evidence to be obtained. The possible courses of action to be taken by the court are set out in s 261(4) to the effect that the court may either give permission to continue with the claim on such terms as the court thinks fit, or the court may refuse permission and dismiss the claim, or the court may adjourn proceedings, giving any directions which it thinks fit.

It is of course possible that such a claim, which it must be remembered is being brought on behalf of the company even if it may be a disgruntled minority shareholder who has in practice commenced the action, will nevertheless be pursued less than diligently by the company's officers if they are in league with the majority shareholders. Just such a contingency is catered for by s 262(2), which provides that a member of the company may apply to the court for permission to continue the claim as a derivative claim. Such an application must demonstrate that the manner in which the company began or continued to claim constitutes an abuse of the process of the court, that the company has failed to prosecute the claim diligently and that it is appropriate for the member to continue the claim as a derivative claim (s 262).

The case law

The old case law on derivative actions – decided before the enactment of the CA 2006 – can be divided between claims based on negligence, claims based on abuse of power and claims based on wrongdoer control. We shall consider each in turn.

That a derivative claim could be brought on grounds of negligence was in itself somewhat controversial. It had once been the position in English law that one could only bring such actions if one could prove fraud. So, in *Pavlides v Jensen* (1956), it was held that no derivative action could be brought when a director failed to get anything like the market value of land belonging to the company when it was sold, because no fraud could be proved. However, Templeman J held in *Daniels v Daniels* (1978) that actions to bring a derivative claim need not be limited to cases of fraud. His Lordship held that 'if minority shareholders can sue if there was fraud, I see no reason why they cannot sue where the action of the majority and directors, though without fraud, confers some benefit on the directors and majority shareholders themselves'. In that case the claimant shareholders sued when directors negligently sold land to one of their fellow directors at a fraction of its true market value. In consequence that director was able to sell the land and make a profit of about £115,000. Clearly this was a level of profit that it would be unacceptable to allow a director to keep. On the facts the director had been able to sell the land for approximately 28 times its purchase price; hence his Lordship's determination that directors should not be able to avoid their liabilities simply on the basis that fraud could not be proven, and instead upheld liability on the basis that they had been merely negligent. As his Lordship put it, it is one thing to have to put up with foolish directors, but it is quite another to have to put up with directors who are so foolish that they make a profit for themselves of about £115,000 at the expense of the company.

There are a number of cases relating to abuse of power. Many of these cases were discussed in the previous chapter in relation to directors' duties. So in *Hogg v Cramphorn* (1967), a mooted takeover of the company was blocked by the inappropriate use of the directors' powers to issue new shares. This was a circumstance that would justify the bringing of a derivative claim. A similar set of facts had arisen in *Menier v Hooper's Telegraph Works* (1873–74), in which the majority shareholders had purported to devolve to themselves the right to divide the company's assets up between the majority shareholders only. It was held by James LJ that it would be a 'shocking thing if that could be done, because if so the majority might divide the whole assets of the company, and pass a resolution that everything must be given to them, and that the minority shall have nothing to do with it'. Evidently then, even if the majority has the power to do something in theory, it will not always be permitted to exercise that power if the exercise of it would be abusive. Presumably these cases and the principle that they propound should now be taken to fall under the heading of default or 'breach of duty' in section 260.

The cases on wrongdoer control are somewhat more confusing. In *Prudential Assurance v Newman Industries (No 2)* (1982), the Court of Appeal considered an application by the plaintiff, who was a large institutional investor in the company holding 3 per cent of its share capital. The plaintiff brought his action against two directors, alleging that they had attempted to defraud the company of £400,000, although this fraud was never proved. The Court of Appeal held that there was no general principle that a derivative action could be brought 'where justice so requires' and that instead the plaintiff must establish a prima facie case. It was held that the rule in *Foss v Harbottle* existed so as to protect the majority shareholders and that plaintiffs were not permitted to bring any action that they pleased, no matter how public-spirited their intentions were.

In *Estmanco v GLC* (1982), Megarry J considered a set of circumstances in which a local authority (GLC) had established a company to manage a large block of flats. The GLC sought to block the sale of those flats to commercial developers, preferring instead to use them for social housing. The issue arose whether or not, because the use of the block of flats was conducted through that company, there had been any wrongdoer control by the majority GLC shareholding in the company in taking this action. His Lordship held that he could see no reason why the minority shareholders should not be entitled to bring a derivative action simply because the majority shareholders considered their action to be in the best interests of the company.

The *Estmanco* case sits a little uneasily with *Prudential Assurance v Newman (No2)* because it seems to be a political judgment brought against the left-wing and somewhat notoriously iconoclastic GLC intended to allow the minority shareholders to countermand the wishes of the majority. The only distinction between the two cases seems to be that *Estmanco* involves a left-wing local authority as majority shareholder, whereas *Prudential Assurance* involves a commercial investor as majority shareholder. If the rule is that the majority shareholder can ultimately control the company, then there is no good reason to decide the cases differently. (A quick read of *Cowan v Scargill* (1985) in trusts law will indicate how Megarry J felt about left-wingers in his courtroom more generally.)

Somewhat unsatisfactorily the law at this time considered wrongdoer control as constituting a fraud on the minority: this requires the courts to embroil themselves in some unedifying debate as to what was meant by fraud in this context. What emerges from these cases is that there need not be any actual fraud, but rather simply some wrongdoing by those in control.

Similarly in *Smith v Croft (No 2)* (1988), it was held that one of the key considerations was whether or not the shareholder was being improperly prevented from bringing a claim on behalf of the company, or whether it was in truth 'the corporate will' and some legitimate organ of the company which was preventing the case being brought. Therefore, if the majority shareholders wanted to ratify the decision in good faith, that would prevent a claim being brought. Other issues which would be considered were the amount of time and cost that would be expended compared to the benefit which might be gained, and the level of inconvenience and managerial disruption which would be caused to the company. This approach tended to narrow, on the case law, the basis on which such an action might be brought by establishing, in effect, a number of hurdles that the claimant would need to cross.

Unfair prejudice to the minority

Unfair prejudice: s 994 CA 2006

The second basis on which the will of the majority shareholders can be resisted is where there has been some unfair prejudice to the minority. The central principle governing the circumstances in which the minority may seek to have some action of the majority set aside on grounds that it is unfairly prejudicial to the minority is set out in section 994 of the Companies Act 2006. That section provides as follows:

> (1) A member of a company may apply to the court by petition for an order under this Part on the ground –
>
> (a) that the company's affairs are being or have been conducted in a manner that is unfairly prejudicial to the interests of members generally or of some part of its members (including at least himself), or
>
> (b) that an actual or proposed act or omission of the company (including an act or omission on its behalf) is or would be so prejudicial.

There are therefore two bases on which such an action may be brought: first, on the basis that the company's affairs have been conducted in a way which is 'unfairly prejudicial' either to the interests of the shareholders in general or to the interests of a particular group of shareholders; alternatively, on the basis that a particular act of the company is similarly unfairly prejudicial. That act may have already been carried out, or it may merely be proposed that it will be carried out in the future, such that any injunction would be appropriate to prevent it being carried out; or the complaint may in fact be based on an omission of the company. The case law has been primarily concerned with the meaning of unfairly prejudicial conduct.

Significantly, the language used in this provision appears to be very broad, as indeed was the predecessor provision s 459 of the Companies Act 1985. However, as will emerge next, the House of Lords took the view that this provision did not give the courts the power (which it may have appeared to give on a literal reading) to decide as they wish: instead, their Lordships sought to impose order on the manner in which the courts would interpret the old s 459.

The leading decision in O'Neill v Phillips

One of the principal questions that underpins most company law is whether the law should stipulate strictly the circumstances in which events should occur, or whether the law should give the courts discretion to decide as they see fit. The finding of an 'unfair' act to the prejudice of the minority shareholders, which is intended to displace an otherwise strict principle that the majority shareholders can do whatever they please within the law, suggests that the court will be deciding by reference to a flexible standard of fairness. This problem of choosing between certainty and flexibility has arisen in the case law on unfair prejudice.

The most significant decision in this area has been that of the House of Lords in *O'Neill v Phillips* (1999), and it is on that leading case which this section will focus primarily. Interestingly, Lord Hoffmann held that while Parliament had intentionally given the court a wide power to do what appears to be just and equitable, that does not mean that the court can do whatever an individual judge considers to be fair. Instead his Lordship held that the concept of fairness must be based on rational principles, but without being explicit as to what those principles were.

In *O'Neill v Phillips*, Mr O'Neill had been a manual worker in the company (Pectel Ltd) since 1983 and had impressed the managing director, Mr Phillips, so much that he had risen through the ranks and eventually been made a director of the company in January 1985 and given 25 shares of the total 100 shares in issue in the company. Phillips was the only other shareholder, with 75 shares. Phillips told O'Neill that he hoped O'Neill would one day take over the running of the company completely, which O'Neill did on 30 December 1985 when Phillips retired. So, in retirement Phillips held 75 per cent of the shares and took 75 per cent of the distributable profits from the company, whereas O'Neill took 25 per cent of the profits in line with his 25 per cent shareholding. However, Phillips did waive his full entitlement so that he and O'Neill would take 50 per cent of the profits each, in spite of Phillips still holding 75 per cent of the shares. This paints a picture of great harmony between the retired owner of the company and the manager of the business. There was a construction boom in the late 1980s from which the company profited greatly. In 1989 and 1990 there was even talk between O'Neill and Phillips that O'Neill would become a 50 per cent shareholder, and advice was taken, but nothing was ever finalised. And then the construction boom ended, just as the company had expanded its business into Germany. Phillips became concerned about O'Neill's management of the business and things began to turn downwards during the recession in the early 1990s. The two men agreed that Phillips would return to manage the business and that O'Neill would move to Germany to manage the business there. It seems that O'Neill's enthusiasm and commitment had waned for some reason, and his performance in Germany was no better than it had been in the UK when the recession began to hit. The two men had a heated argument on 4 November 1991 about O'Neill's performance in Germany.

This is a common situation in private companies. Strong personal relationships are formed and enmities can also develop. These sorts of human dramas will of course impact directly on the performance of the business, even though the company structure may not seem to change. Who knows what personal dramas could impact on performance at work of any corporate executive? Over a number of years, close friendship can change to disillusion and even dislike. A star is selected from the building site, entrusted in time with running the business, and then his star fades until he is in a heated argument with his mentor that brings their relationship to an end. To return to the case, O'Neill severed his links with the company and set up a competing business in Germany.

In *O'Neill v Phillips* the litigation began initially in correspondence between the parties as to whether Phillips had broken his word to O'Neill by failing to make him a payment of 50 per cent of the profits, whether there had been dishonesty by Phillips, and whether O'Neill could sever the 'partnership' between the parties. Importantly, of course, this was not a 'partnership' but a shareholding in a company in which the claimant was also a director. Nevertheless, it is significant that the argument was put in terms of there being a sort of partnership between the two men who were also the only shareholders of the company, because this has become a feature of some of the case law, such as *Re RA Noble & Sons (Clothing) Ltd* (1983) where exclusion from a company which had originally been formed as a quasi-partnership between its founders was found to be unfairly prejudicial conduct. It is a feature of this sort of litigation, even among commercial people, that it begins with acrimony and a failure to identify at the outset the features that will propel the case over eight years from initial correspondence between solicitors to the decision in the House of Lords. It is interesting that these two men spent as much time embroiled in litigation as they had spent working together in the company.

The precise legal issue in *O'Neill v Phillips* was whether Phillips, as the majority shareholder (with 75 per cent of the shares), had acted in a way which was unfairly prejudicial to the minority shareholder O'Neill (with 25 per cent of the shares), such that O'Neill should be entitled to an order under (what is now) s 994 CA 2006. Having summarised the facts and the decisions in the lower courts, Lord Hoffmann began his judgment on the interpretation of what is now s 994 with the following words:

> Parliament has chosen fairness as the criterion by which the court must decide whether it has jurisdiction to grant relief. It is clear from the legislative history (which I discussed in *Re Saul D Harrison & Sons plc* (1995) at 17–20) that it chose this concept to free the court from technical considerations of legal right and to confer a wide power to do what appeared just and equitable. But this does not mean that the court can do whatever the individual judge happens to think fair. The concept of fairness must be applied judicially and the content which it is given by the courts must be based upon rational principles. As Warner J said in *Re J E Cade & Son Ltd* (1992)

at 227: 'The court . . . has a very wide discretion, but it does not sit under a palm tree.' (at p.966)

This passage is fascinating. What is clear is that the statutory language (which has led now to s 994) intended to free Parliament from 'technical considerations', and yet Lord Hoffmann is determined that this is not to be read as granting freedom to the courts. Instead, Lord Hoffmann takes an approach which is typical of English judges when they are confronted with a statutory provision that appears to grant them a wide licence to do whatever they think appropriate, which is to impose constraints on the way in which that statutory power should be used in practice. (The psychologist Erich Fromm (1942) writes about the human fear of being completely free because for most people that means a lack of structure or dependability: it always seems to me that there is something of that fear in the judges when they behave in this way. Family law is perhaps the only area of English law that does take such an open-textured approach to its principles.) So English judges are not permitted to decide on the basis of fairness: instead they must base their judgments on 'rational principles'.

Lord Hoffmann continued:

> Although fairness is a notion which can be applied to all kinds of activities, its content will depend upon the context in which it is being used. Conduct which is perfectly fair between competing businessmen may not be fair between members of a family. In some sports it may require, at best, observance of the rules, in others ('it's not cricket') it may be unfair in some circumstances to take advantage of them. All is said to be fair in love and war. So the context and background are very important.

What this passage suggests is that judges are not considered capable of developing a notion of 'fairness' which is appropriate only to cases involving commercial parties ('businessmen') and distinct from what we might consider appropriate in a person's private life. Therefore, even though Parliament has specified a notion of what is 'unfair' in the statute, Lord Hoffmann does not wish to engage with what that word might mean. The reference to sport is more obscure still. Lord Hoffmann seems to suggest that some sportsmen (by which I mean both genders) might observe the rules literally but still act in a way which we might consider 'unfair'. This might mean in football ignoring the convention that when the ball is put out of play because a player is badly injured, the ball is returned to the team formerly in possession at the resulting throw-in. An Arsenal player in 1999, ignorant of this convention, once took advantage from the throw-in and scored. This led to the match being replayed because the goal was considered unfair. What this example shows is that on a case-by-case basis it is possible to identify whenever a convention of this sort has been breached. Indeed, referees are constantly required to interpret the application of the rules of the game as a match progresses: the current offside rule, with its nebulous concept of 'interfering with play' is a good example. What also emerges is a warning that Lord Hoffmann might be unreliable in his personal relationships because he seems happy with the idea that 'all is fair in love and war', instead of there being a possibility of judging the fairness of a person's behaviour in a love affair! What emerges is that Lord Hoffmann does not want there to be too much flexibility here.

His Lordship accepted that:

> company law has developed seamlessly from the law of partnership, which was treated by equity, like the Roman societas, as a contract of good faith. One of the traditional roles of equity, as a separate jurisdiction, was to restrain the exercise of strict legal rights in certain relationships in which it considered that this would be contrary to good faith. These principles have, with appropriate modification, been carried over into company law.

Nevertheless, Lord Hoffmann saw much merit in being strict in the application of these principles. His Lordship made reference to the decision of Lord Wilberforce in *Ebrahimi* (discussed below), which had followed a more flexible and equitable approach, but disposed of it on the following basis:

> In my view, a balance has to be struck between the breadth of the discretion given to the court and the principle of legal certainty. Petitions under s [994] are often lengthy and expensive. It is highly desirable that lawyers should be able to advise their clients whether or not a petition is likely to succeed. Lord Wilberforce, after the passage which I have quoted, said that it would be impossible 'and wholly undesirable' to define the circumstances in which the application of equitable principles might make it unjust, or inequitable (or unfair) for a party to insist on legal rights or to exercise them in a particular way. This of course is right. But that does not mean that there are no principles by which those circumstances may be identified. The way in which such equitable principles operate is tolerably well settled and in my view it would be wrong to abandon them in favour of some wholly indefinite notion of fairness.

So, in spite of the fact that Lord Hoffmann recognised that company law developed from partnership law and trusts law, and that one of the traditional roles of equity in this context was to restrain the exercise of strict legal rights where to do otherwise would be contrary to good faith, and furthermore that those principles have been carried over into company law, nevertheless his Lordship refused to accept that the concept of fairness in the legislation should be interpreted in an equitable fashion.

What is more remarkable is that, while Lord Hoffmann claims that the principles are 'tolerably well settled', we should recall that his Lordship had to recognise that his own attempt to settle those principles in *Saul D Harrison* should now be abandoned: which leaves us wondering exactly how they can be said to well settled if the previous authority is now to be ignored. In *Saul D Harrison*,

Hoffmann J held that the principles should be based on deciding whether or not the claimants had a 'legitimate expectation', like that in public law, that they would be treated in a particular way: an approach which he now abandoned. Lord Hoffmann referred to the decision of Parker J in *Re Astec plc* (1998), where Parker J referred explicitly to the notion of legitimate expectation needing to be qualified by asking whether or not the conscience of the defendant had been affected. Oddly, in spite of the continuing role of the notion of conscience in modern trusts law (see *Westdeutsche Landesbank v Islington* (1996)), Lord Hoffmann treated it as though it was a relic of a bygone age.

This is all most unsatisfactory given that Lord Hoffmann was very vague about what he thought that the more rigid rules that he preferred would actually be. What Lord Hoffmann was really concerned about was the cost of these applications and the need for lawyers to be able to give clear advice to their clients during litigation, rather than leaving matters up to the discretion of the court (even though that was what Parliament intended). That Lord Hoffmann did not make it clear what he intended the certain principles to be means that the lawyer giving advice in practice is in a no clearer position.

So, despite citing the decision of the House of Lords in *Ebrahimi v Westbourne Galleries* (1973), where a more traditionally equitable approach to this concept was taken, Lord Hoffmann nevertheless refused to treat it as an appropriate concept to be used to decide these cases. In *Ebrahimi*, Lord Wilberforce had expressly suggested that it would be inappropriate to delimit the basis on which that concept might operate. (It could be said, of course, that Lord Wilberforce had been focusing primarily on a different statutory provision; however, it was Lord Hoffmann himself who drew the parallel between the two cases.) What is unfortunate about the decision in *O'Neill* is that while Lord Hoffmann does appear to accept that these are quasi-equitable principles, he nevertheless fails to deal with them in an equitable way.

The concept of 'unfair prejudice' in the case law

An early House of Lords decision in *Scottish Co-operative Wholesale Society Ltd v Meyer* (1959) centred its focus on whether or not the affairs of the company had been conducted in a way that was 'oppressive' to the applicants. However, a progressive change in the legislation has meant that our focus is now on 'unfairness' rather than 'oppression', which suggests a milder level of harm being caused to the applicants. What can be gleaned from the decision of the House of Lords in *O'Neill v Phillips* is that the concept of unfair prejudice is not to be too tightly drawn, but yet Lord Hoffmann wants it to be developed in accordance with rational principles. In New Zealand it has been held that a decision as to whether or not there has been unfair prejudice will require a balancing of the conflicting interests of the various participants in a company (*Thomas v HW Thomas Ltd* (1984)).

A clear example of unfair prejudice is where a controlling director or shareholder allots themselves shares so as to diminish the voting power of other shareholders. In *Re DR Chemicals Ltd* (1989) it was held that it was a clear case of unfair prejudice for a director and shareholder of a company to allot himself enough shares to take his shareholding from 60 per cent to 96 per cent of the total shareholding. Similarly, in *Dalby v Bodilly* (2004) there was unfair prejudice when enough shares were issued to dilute one shareholder's holding from 50 per cent to only 5 per cent of the total shares in issue.

Another example of unfairly prejudicial conduct in the decided cases falls under the general heading of diverting company property to particular shareholders. Company assets in this context will include business opportunities and intellectual property as well as money or chattels. So in *Re London School of Electronics Ltd* (1986), it was held to be unfairly prejudicial for the defendant to divert students away from the company to a new venture which the defendant was setting up.

By contrast, there will not be unfair prejudice simply because of the late presentation of accounts (*Ringtower Holdings Plc* (1989)). In *Marchday Group Plc* (1998) Neuberger J held that if there was no breach of the company's articles of association and nothing that was unlawful, there could not be a petition for unfair prejudice.

A frequent cause for complaint is the mismanagement of the company. Whether that will constitute unfair prejudice to the shareholders due to the effect on the value of their shareholding will depend upon the seriousness of the mismanagement (*Re Elgindata Ltd* (1991)). It has been held that a serious abuse of power or a serious ulterior motive would be needed to found a successful petition (*Re Saul D Harrison* (1995)); although Arden J has been prepared to accept mismanagement that would cause economic loss to the company would support an application under what is now s 994 (*Re Macro Ipswich Ltd* (1994), where the directors had failed to inspect and maintain the buildings owned by a property company so that the company had lost money). It has also been held to be unfairly prejudicial to take an excessive level of remuneration from a company (*Irvine v Irvine* (2006)). Exclusion from the management of the company has been held, in small companies in a quasi-partnership context, to constitute unfair prejudice in *RA Noble & Sons* (1983) and *Richards v Lundy* (2001).

Winding up on the 'just and equitable' ground

Section 122(1)(g) and the decision in Ebrahimi v Westbourne Galleries

The third basis on which the will of the majority shareholders can be resisted is where the company is wound up on the 'just and equitable ground' in the insolvency legislation. What this means is that the company is brought to an end: it is 'wound up'. Clearly, the company must be in a parlous condition, or the relationships between its participants must have deteriorated to a remarkable extent, to justify bringing the company to an end; and this would include circumstances in which the minority shareholders are trapped in a company where they are suffering in some way, and where the rule in *Foss v Harbottle* means that they have no power to extricate themselves (and in relation to private companies where no buyer could be found, no power to remove their capital from it). Therefore, having the company wound up might be the only way to allow the minority shareholders to extricate themselves and their capital from the company, and to allow all of the participants to divide up the company's property and move on. With this in mind, it is provided by s 122(1)(g) of the Insolvency Act 1986 that:

A company may be wound up by the court if . . .

(g) the court is of the opinion that it is just and equitable that the company should be wound up.

Therefore, there is an ostensibly broad power for the court to wind up companies. Having considered the principles relating to 'unfair prejudice' and the limitation that the House of Lords sought to place on that principle, the question arises whether there would be a similar limitation in relation to s 122(1)(g).

The best illustration of this principle in action is in the leading case of *Ebrahimi v Westbourne Galleries* (1973), where a company had been formed by two men effectively as partners. This concept of quasi-partnership was mentioned briefly above. In essence, what is meant is the following. Companies are distinct legal persons with all of the bureaucratic requirements discussed in this book; whereas partnerships can be informal arrangements between people that are based on a contract and an undertaking in common with a view to profit, but where the partnership has no legal personality itself. The reference to 'quasi-partnerships' is to situations in which a company is formed but in circumstances in which a small number of people (typically two) have created a company as the vehicle for the small business that they began together commercially as a joint undertaking between them. In spite of the formalities of company law, the courts do seem to have become dewy-eyed about these sorts of situations when things deteriorate and the business collapses acrimoniously. (Other examples of small companies which were treated as quasi-partnerships with similar results are *Re Ghyll Beck Driving Range* (1993) and *R&H Electrical Ltd v Haden Bill Electrical Ltd* (1995).)

In *Ebrahimi*, the founders of the business shared the management of the business and all the profits; they were also the original subscribers to the memorandum of association when the company was formed, and they were its first directors. It was these factors that led the court to hold that it was a quasi-partnership. In time a third man was made a director of the company and granted shares in the company: this third man was son to one of the original founders of the company. Consequently, the father and son together had enough votes to constitute a majority shareholding in the company. Over time the father and son fell out with the other director and effectively ousted that other director from the company. As a result, that minority shareholder and director decided to present a petition to the court seeking the winding up of the company.

Lord Wilberforce gave the leading speech in the House of Lords. Unlike the judgment of Lord Hoffmann considered above, Lord Wilberforce appeared to be much more comfortable with the open-textured language of the statute: to whit, the requirement that it be just and equitable to wind up the company.

> The foundation of it all lies in the words 'just and equitable' and, if there is any respect in which some of the cases may be open to criticism, it is that the courts may sometimes have been too timorous in giving them full force. The words are a recognition of the fact that a limited company is more than a mere legal entity, with a personality in law of its own: that there is room in company law for recognition of the fact that behind it, or amongst it, there are individuals, with rights, expectations and obligations inter se which are not necessarily submerged in the company structure. That structure is defined by the Companies Act and by the articles of association by which shareholders agree to be bound. In most companies and in most contexts, this definition is sufficient and exhaustive, equally so whether the company is large or small. The 'just and equitable' provision does not, as the respondents suggest, entitle one party to disregard the obligation he assumes by entering a company, nor the court to dispense him from it. It does, as equity always does, enable the court to subject the exercise of legal rights to equitable considerations; considerations, that is, of a personal character arising between one individual and another, which may make it unjust, or inequitable, to insist on legal rights, or to exercise them in a particular way.

Here we see a sensitivity to the dissentient ideas considered in Chapter 2 to the effect that there is more to a company than distinct legal personality and the blank facade of the instrument of incorporation. Instead, there are human beings – with their aspirations, hopes and dreams – inside companies. Therefore, the courts should be prepared to delve into the circumstances and decide whether the interaction between those human beings has reached such a pitch that the company should be wound up so that its property can be distributed among the shareholders.

Furthermore, Lord Wilberforce was very clear that he did not wish to delineate the circumstances in which the statute should be used. He considered that to be wholly undesirable:

> It would be impossible, and wholly undesirable, to define the circumstances in which these considerations may arise. Certainly the fact that a company is a small one, or a private company, is not enough. There are very many of these where the association is a purely commercial one, of which it can safely be said that the basis of association is adequately and exhaustively laid down in the articles. The superimposition of equitable considerations requires something more, which typically may include one, or probably more, of the following elements: (i) an association formed or continued on the basis of a personal relationship, involving mutual confidence – this element will often be found where a pre-existing partnership has been converted into a limited

company; (ii) an agreement, or understanding, that all, or some (for there may be 'sleeping' members), of the shareholders shall participate in the conduct of the business; (iii) restriction upon the transfer of the members' interest in the company – so that if confidence is lost, or one member is removed from management, he cannot take out his stake and go elsewhere.

He was clearly sensitive to the fact that this was a case involving a quasi-partnership, particularly when it was held that 'a winding up may be ordered if such facts are shown as could justify a dissolution of partnership between them'. On the facts Lord Wilberforce was impressed with the fact that the claimant was being treated as a mere employee and not as a director and that his rights as a minority shareholder were being effectively overlooked. Therefore, an order to wind up the company was made.

A way forward

It is interesting to compare the approaches of Lord Wilberforce in *Ebrahmi* and Lord Hoffmann in *O'Neill*. The latter appears determined to treat company law as being a branch of commercial law more generally, requiring strict legal rules developed on rational principles. This is so even when there is legislation in place expressed in terms of much more general principles. Interestingly, but perhaps not surprisingly, this is another example of judges thinking they know better than Parliament. By contrast, Lord Wilberforce almost relishes the equitable nature of the principles in the statute. His Lordship was clear that we must not limit those principles and their future operation by defining them too tightly. It is natural, perhaps, in our commercial life to wish that we could create legal rules which always generate the correct result and which are always certain and predictable in their operation. However, it is the way of the world that any attempts to create comprehensive programmes of strict legislation always come unstuck when confronted by the normal chaos of ordinary life, even among commercial people. In such circumstances, the courts must accept that broad principles serve us every bit as well as strict legal rules; hence the inclusion of the concept of the just and equitable ground in the 1986 Act, precisely because it gives the courts some flexibility in the application of the law to difficult factual circumstances. Furthermore, of course, not all companies are trading companies in any event, and a flexible principle will therefore enable the right result to be achieved across the full range of cases while using the same law.

Shareholder powers

The power to remove directors

One of the principal powers of shareholders in UK company law is the power to remove directors under s 168 CA 2006. To this end, it is provided in s 168(1) that:

a company may by ordinary resolution at a meeting remove a director before the expiration of his period of office.

Therefore, one of the principal ways in which shareholders can control the directors is with this threat of removal. Importantly, nothing in any agreement between the company and the individual director can prevent this power from being exercised if the appropriate majority of shareholders votes for it (s 168(1)). The ordinary resolution at the company meeting requires simply therefore that more than half of the members agree to remove the director in question. If there is a resolution being put before a meeting to remove a director, or to appoint somebody else as director in her place, special notice must be given of that resolution (s 168(2)). In essence, this gives both members and directors an opportunity to marshal themselves in advance of that meeting as opposed to having this issue sprung on them as a surprise at a general meeting of the company. If the director is removed, the vacancy so created can be filled as a casual vacancy (s 168(3)).

Section 169 CA 2006 provides the director in question with a right to protest against the resolution suggesting her removal. The director is entitled to receive a copy of the notice relating to that resolution. Importantly, the director is entitled to be heard at the meeting when the resolution is discussed (s 169(2)). The director is entitled to make representations in writing. While copies of these representations do not need to be circulated, the court must be satisfied that the rights conferred by s 169 are not being abused.

Practical power and the composition of shareholders as a body

This chapter has focused on the law relating to shareholder democracy. What has emerged, though, is that it is a form of democracy based on the size of a person's investment and the concomitant number of shares acquired, but it is not based on 'one person, one vote'. There is another significant aspect to shareholder democracy: that of naked power. In public companies the principal shareholders are institutional investors (as discussed in Chapter 3) and their power and influence are of a different order from other shareholders. Whereas none of those institutional shareholders individually would be likely to hold a majority shareholding in a public company, as a group the institutional investors will be able to acquire access to senior management and be able to process the information that they are given in entirely different ways from private individuals. Furthermore, the expectations of the institutional investor community as a group – which is what drives share prices in public companies more than anything – have considerably greater influence on the company's directors than the interests of any other class of shareholder. Even the occasional newsworthy demonstration by groups of private shareholders on the day of an annual general meeting does not have anything like the impact of the day to day influence of institutional investors through their relationship managers and their analysts.

One of the key ideas which drives the management of public companies is the business school concept of 'earnings per share growth', which leads institutional investors to demand that the amount of profit earned in a financial year per share in issue must increase year-on-year for that company to be considered to be a success. So, while each share has the same rights in law, the power that institutional investors have is of a different order because of their expertise and the significance of their investments as a group in the company.

Moving on ...

In this chapter we have considered the phenomenon of shareholder democracy and the circumstances in which statute protects the rights of minority shareholders. In the next chapter we move on to consider how non-legal codes of practice are used in companies to protect the best interests of the company and of the shareholders through good corporate governance practice.

Chapter 9

Corporate governance

Introduction

Corporate governance is concerned with practical arrangements that seek to ensure that executive directors are not able to abuse their legal powers or any other influence that they may have within their organisations. Consequently, corporate governance norms purport to regulate the day to day practice of corporate management. This chapter considers the Combined Code on Corporate Governance, which is the applicable code for public companies in the UK. It is not legally binding. Instead, such companies are intended to choose to conduct their internal processes in accordance with it, and the listing rules (within securities regulation, discussed in Chapter 11) require such companies to state whether or not they agree to be bound by it. We shall consider the key provisions of this code before considering some of the literature that has analysed the development of corporate governance norms in a variety of jurisdictions.

Finally, as a counterpoint to all of this non-binding 'soft' law, we shall consider the legislation dealing with the disqualification of directors from office. It is suggested that it is legislation of this sort that removes directors which has considerably more direct effect on the behaviour and treatment of directors than any non-binding legal code. The point, then, is that corporate governance norms in the UK are in fact just another messy compromise between legislators and capitalists and that the legislators will not interfere with the freedom of the capitalists to do whatever they want to enrich us all.

The impetus to introduce greater control on the management of large public companies in the USA came in the wake of the scandals at Enron, WorldCom and others, and the corporate governance failures in huge investment banks such as Lehman Bros where almost unimaginable risks were taken in markets that subsequently collapsed. Before those scandals broke, there had already been concern in the UK about levels of executive pay and lack of practical controls on the power of directors in large public companies (particularly when they already had the majority institutional shareholders on their side). There had always been pressure against formal corporate governance rules in the UK on the basis that the freedom of entrepreneurial company directors should not be constrained by 'bureaucrats', or 'the dead hand of government', or 'the nanny state'. Therefore, the following compromise was reached in the UK: a code of standards of good behaviour for the management of public companies would be created by an independent process and, importantly, it would not be legally binding.

Corporate governance in the UK

The context

Corporate governance within company law

Corporate governance has become an increasingly important topic within the study of company law. Strictly speaking it is not actually a part of company law, but rather constitutes a collection of principles which public companies in particular are expected to observe in the operation of their businesses. Corporate governance focuses mainly on the behaviour of senior management. In the UK, corporate governance thinking emerged originally out of concerns about three principal issues: the level of directors' remuneration, the reliability of the audit of company accounts and fears about the unbridled power of chief executive officers and small cabals within senior management. Therefore, the earliest steps towards comprehensive corporate governance in the UK focused primarily on these issues. However, rather than have government introduce legislation to deal with these matters, a consensus was reached whereby formal reports were published by various committees investigating corporate governance practices, which the largest public companies (by and large) agreed would be binding on them in practice, if not in law.

The competing positions

There are two ostensibly opposing views on the current organisation of capitalism in the world: either we should allow capitalists and the directors of trading companies complete liberty in free markets to drive our economies onwards, or we should constrain the actions of capitalists and the directors of trading companies to prevent them from acting unconscionably in a single-minded search for profit, which will concomitantly have a severe impact on our societies and on the environment. During the late 1980s and the 1990s there was a drift towards light-touch regulation of financial and corporate markets generally; many would blame the calamitous global financial crisis of 2007–10 on that attitude to regulating financial markets. An ongoing concern about executive pay in the UK prompted the argument in the late 1980s that the governance of large public companies required some control. Any suggestion that legislation should be introduced was resisted on the basis that it would constrain the freedom of company directors to achieve growth for their companies and thus for the economy. Consequently, a muddy compromise of non-binding, non-legal codes of corporate governance practice was developed, although those principles have tended to be important in practice.

The large corporate scandals that brought corporate governance onto the agenda

The impetus to introduce stronger corporate governance codes in the UK and to begin the process in earnest in the USA was the result of a series of corporate scandals. The UK was ahead of the USA in this regard, until the corporate scandals broke at Enron and WorldCom. It was these huge corporate collapses, caused by the falsification of company accounts by secretive groups of senior management, which focused attention on the need for greater corporate governance in the USA.

Many of the largest corporate scandals and scandals involving financial institutions have involved an ogre at the centre of the organisation who inspired such fear or reverence in other employees (and fellow directors) that he was able to drive the organisation into the ground: for example, Dick Fuld at Lehman Bros, Robert Maxwell at the Mirror Group, Jeff Skilling at Enron and Bernie Ebbers at WorldCom. Better management systems, in which outsiders could have influenced senior executives or held them to account, might have prevented those abusive management practices from developing in the first place. Management systems also need to deal with lower-level employees so as to identify the sorts of fraud committed by people like Joseph Jett at Kidder Peabody, who was described as physically intimidating fellow workers so that no one questioned the unusual levels of profit he was earning by hiding his losses. More famously the medium-level trader Nick Leeson (immortalised in the film *Rogue Trader*) bankrupted Barings Bank by concealing his enormous losses in a fraudulent account and then booked those losses as enormous projected profits. He avoided telephone calls from his bosses in London when they called him in Singapore to question his profits. There was a complete failure on the part of management to put proper systems in place and to understand the derivatives business that their employees were running.

The Enron collapse

Perhaps the most egregious collapse of a large corporation due to a lack of proper corporate governance was that of the short-lived American behemoth Enron. The book *The Smartest Guys in the Room* by Bethany McLean and Peter Elkind (2004), and the subsequent documentary film of the same name, detail the way in which the corporate culture at Enron allowed a small coterie of senior executives to commit fraud and to bankrupt their company. It was a failure of corporate governance that led to what was the largest corporate failure in the USA, until the collapse of Lehman Brothers in 2008.

The Enron story has two key elements of classical Greek drama: hubris and deceit. Enron moved from a traditional business supplying natural gas to acting as a supplier of electricity and as a sort of exchange for other people who held or who needed energy in the future. This became a market for financial instruments known as 'energy derivatives' in which people would buy and sell the right to sell or receive energy in the future. Enron traded these instruments in a way that appeared to the outside world to be remarkably successful. Together with a move-ment to 'mark-to-market' accounting, Enron was able to trade in financial instru-ments and to record the value of those instruments in its accounts at whatever they considered their market value to be from time to time.

Mixed with this development of a new industry, the corporate culture at Enron was legendarily cut-throat. A macho culture spread among its traders, as is often the case on financial trading floors. This led to a culture of risk-taking within the company. A small coterie of senior executives became all-powerful within the organisation, with the result that their decisions were beyond question within the company. In practice what this meant was that Enron could pretend to be able to maintain perpetual profits by massaging their accounts: in essence, future trans-actions were recorded as immediate profits. The journalist Bethany McLean wrote an article in March 2001 that asked how Enron was making money when in fact it had no cash coming into the business, even though it was posting enormous profits.

A cash flow analysis of Enron showed that it was not receiving money, even though it recorded as profits moneys that it hoped it would receive in the future. The commodities derivatives trading which it had shifted into was being used to generate phantom profits, while its natural gas business was losing money rapidly and, for example, its mooted video-on-demand business did not make a penny but had profits of US$53 million recorded in the accounts. Enron milked the deregu-lated Californian energy market criminally – causing rolling blackouts across the state – to maintain its ostensible profitability. Senior executives sold hundreds of millions of dollars worth of stock options as they realised that the company was in trouble. The company was indulging in large-scale accounting fraud. It was that simple. Debts of US$30 billion were hidden, in part by entering into bogus transac-tions with special purpose vehicles to hide the debt away and help to make it appear as though Enron itself was profitable. Together with a bubble of false profits on derivatives transactions, this hid the problems from the public view. When the business collapsed on 2 December 2001, having been valued at US$60 billion not long before, it did so remarkably quickly. A company like Enron, which was built on a froth of financial instruments, would quickly collapse when confidence in the company disappeared. McLean's article was one of the pins that burst that bubble.

Before the collapse, Enron traders had been involved in criminal trading and other executives in accounting malpractice. Stories of strippers on the trading floor, macho weekends in the mountains, and so forth, are recorded by Fox (2002) and by McLean and Elkind (2004) as part of the company's culture. It is important to note that, in these sorts of atmospheres, it is possible for common sense to be subsumed within the corporate culture. The senior executives were focused on generating ostensible profit growth so that the share price would continue to rise. Good corporate governance practice should act as a bulwark against these sorts of events. The corporate governance point was that senior executives had been able to falsify the accounts, so that it appeared that Enron's movement into commodi-ties and derivatives trading was generating enormous profit when in fact it was losing money, without anyone asking questions.

In time, the auditors Arthur Andersen became complicit in the false accounting at Enron, and their fault as a firm caused Arthur Andersen to go into bankruptcy. Andersen is estimated to have received US$1 million per week in fees from Enron. When the clouds began to gather, Andersen shredded one ton of files it held on Enron. A large number of American investment bankers have been alleged to have been involved in dealing with Enron's off-balance-sheet transactions. This sort of complicity between companies and their advisors demonstrates how important it is to have robust corporate governance procedures to ensure that cosy deals where auditors approve fraudulent accounts, or where bankers enter into asset transactions to hide assets from a company's balance sheet, cannot take place. (You are strongly advised to watch the DVD, which remains easily obtainable. The similarly enormous collapse of WorldCom is well documented by Lynne Jeter in *Disconnected – Deceit and Betrayal at WorldCom* (2003).)

Thus, companies need robust internal controls to ensure that fraud or other abuses are not being committed. With this is mind, we shall consider the Combined Code on Corporate Governance.

The Combined Code on Corporate Governance

The underpinning purpose of the Combined Code

In the UK the principal reservoir of rules dealing with corporate governance in public companies is contained in the 'Combined Code on Corporate Governance', published in June 2006 (the Combined Code). The Combined Code applies to public companies, and it is a requirement of securities regulation that such companies must include a statement in their annual report as to whether or not they are complying with the code (Listing Rules, 9.8.6(5)R). That aside, this code is not legally binding. Instead it is something that is known as 'soft law', which means that while not being legally enforceable principles, it is hoped that the code may affect the way in which people behave. The idea is that, if a public company failed to implement the corporate governance standards in their jurisdiction, this would be a signal to investors, to the company's bankers and to the public generally that the company is poorly managed and is thus unreliable.

A key part of a company's reputation is the manner in which it is managed or perceived to be managed. Therefore, letting it be known that it complies with those norms can be an important part of a company's reputation. Nevertheless, the effectiveness of this sort of soft law is difficult to determine. We shall return to the justifications for this sort of principle later. In the meantime, we shall focus on the Code itself.

To make a long story short, one of the principal concerns about the way in which public companies were being managed was that their executives had unfettered control. Therefore, it was considered necessary to involve non-executive directors in the decision-making processes of the board in specific contexts relating to the activities of directors, the audit process, the setting of directors' remuneration levels and the company's relations with its shareholders. Each is considered in turn.

The underlying principle

The underlying objective of these principles is to balance out executive power with non-executive power. This means balancing out the power of the full-time executives with other board members who do not have an executive role within in the company, but who instead operate as outsiders who can check the power of the full-time executives, especially in relation to particularly sensitive areas of the board's activities. In large part this required balancing out the power of the chief executive officer (CEO) with a non-executive chairman, where the former was in control of the day to day management of the business and the latter was responsible for advising and where necessary checking the CEO.

The Code creates checks and balances for executive power while also placing responsibility for some sensitive aspects of decision-making with non-executive directors. Clearly, the abilities and experience of these non-executive directors are particularly important in the operation of this system. There has been some concern that the non-executive directors of large public companies are in practice drawn from groups of people who are already senior executive directors of other large public companies, so that we might wonder whether or not their much-vaunted independence is in fact tempered by a tendency to reach the same decisions as the executive directors would reach in any event.

Principles relating to directors

The core principle of the code relating to directors is that if the directors' means of conducting business are inadequate, or if their personal qualifications or knowledge as individuals are inadequate, that will evidently have an impact on the management of the company. Effective management is clearly beneficial for a company and it is something that is of enormous importance to potential investors. The famous investor Warren Buffet of Berkshire Hathaway always stresses the significance of dependable, skilful management in the success of a company. Therefore, it is one of the assumptions underpinning the principles of corporate governance that the more knowledgeable directors are and the more those directors observe minimum standards of behaviour in the operation of their companies, the better that company will be.

The main principles in the code relating to the board of directors are that 'every company should be headed by an effective board, which is collectively responsible for the success of the company' (Combined Code, A.1). The board is expected to provide entrepreneurial leadership for the company while also acting 'within a framework of prudent and effective controls'. Any company must straddle the need to take risks so as to earn profit with the need for sufficient prudence when managing the company. This obligation could be understood as operating in parallel with the duties of directors discussed in Chapter 7 to promote the success of the company (CA 2006, s 172). Thus, the fiduciary duties of directors require that they both act in the best interests in the company and also that they balance the need to take risks with prudence in the management of the company.

One of the most important principles of corporate governance is the establishment of the role of chairman, who carries on some of the senior management responsibility in the company and who balances out the power of the chief executive officer, who has day-to-day control of the company's activities (Combined Code, A.2.1). The Combined Code provides that:

> There should be a clear division of responsibilities at the head of the company between the running of the board and the executive responsibility for the running of the company's business. No one individual should have unfettered powers of decision (Combined Code, A.2.1).

The chairman will chair sensitive committees in a non-executive capacity so that the CEO is not constrained in the day-to-day operation of the business, but also so that the CEO is not able to drive the company off on an undesirable path, nor to run the company in an undesirable way over the longer term. To ensure that the entire board of executive directors is not able to lead the company astray, 'non-executive directors' are required to be appointed to the board of directors (Combined Code, A.3). Again, the aim is not to wrest control of the business away from the executives, but rather to ensure that they do not have complete control over sensitive aspects of the company's activities – such as setting the directors' own remuneration or approving the annual accounts – without being called to account. Therefore, as will emerge below, the Code requires that there are committees created for many of these functions that are staffed either wholly or in part by non-executive directors. The chairman is expected to meet regularly with the non-executive directors without the executive directors being present; and the non-executive directors are also expected to meet to review the performance of the chairman.

It is important also that appropriate people are appointed to the board. Because it is generally considered that the board of directors of a public company should have vision, integrity and the strength of character to question the company's strategy as well as to pursue it, then the appointment of new directors to the board is consequently required to be 'formal, rigorous and transparent' (Combined Code, A.4). The company therefore needs to develop appropriate procedures for the conduct of this aspect of corporate governance. Being a director is not in itself a profession and therefore new directors may not necessarily know the obligations that company law and corporate governance principles place on them. When new directors join the board, they should be inducted and trained appropriately. The obligations of directors in listed companies, for example, are particularly onerous (Combined Code, A.5); this is especially so in relation to the preparation of prospectuses when securities are issued and the continuing obligations of disclosure in relation to prospectuses which are imposed by securities regulations (as discussed in Chapter 11). The board of directors is expected to 'undertake a formal and rigorous annual evaluation of its own performance and that of its committees and individual directors' (Combined Code, A.6). The directors are expected to be open to regular re-election by shareholders with a view to the regular refreshing of the board (Combined Code, A.7).

Directors' remuneration

One of the key prompts behind the creation of the Combined Code was the rapidly increasing level of executive pay in public companies, such that the differential between the pay of senior management and the pay of junior workers was widening rapidly. It is commonly argued that directors need to be well paid so as to attract the best candidates. However, many investors are worried that directors pay themselves too much, with the result that this bleeds cash out of the company and distorts their incentives for working for the company. Consequently, the Code embodies an attempt to find a middle way between the imposition of direct controls on executive pay and allowing the company to pay large amounts to directors to attract the highest calibre of person.

The central principle in the Code is to 'attract, retain and motivate directors of the quality required to run the company successfully' and yet not to pay 'more than is necessary' (Combined Code, B.1). The Code provides that a company's policy on executive remuneration should be developed by means of a 'formal and transparent' procedure, which is to be conducted by a remuneration committee in consultation with the Chairman and the CEO (Combined Code, B.2). The remuneration committee should comprise three non-executive directors to provide independence in remuneration policy.

Accountability and audit

Given the history of many of the largest corporate scandals in the UK and the USA relating to fraudulent accounting, perhaps the most significant part of the Code relates to the preparation of the accounts and the auditing function more generally. It is also significant that the directors make themselves accountable to the shareholders more generally. In essence, the Code is concerned in this sense to submit the actions of the directors and the performance of the company to the disinfectant properties of daylight: that is, to make everything visible to the shareholders, and thus to enhance 'transparency' in the public policy jargon. To combat abuse in this context, the aim of the Code is to ensure that the auditing of companies' accounts is scrutinised by an audit committee made up at least in part of non-executive directors.

There are three ways in which the board of directors makes itself accountable to the shareholders for the performance of the company: by financial reporting throughout the financial year, by the maintenance of internal controls to monitor the company's performance in terms of the position of shareholders, and by the effective audit of the company. The provisions of company law relating to the preparation of accounts and the audit of those accounts were considered in Chapter 6. In relation to financial reporting of the company's position the general principle in the Combined Code is that the 'board should present a balanced and understandable assessment of the company's position and prospects' (Combined Code, C.1). This principle governs not only annual accounts but also interim financial reports, which are significant in relation to the listing rules, and any publication in relation to price-sensitive information. Second, there is also a need for the development of internal controls that are sufficiently sound 'to safeguard shareholders' investment and the company's assets'. The board is expected to conduct an annual review of these internal controls and to report to the shareholders as to the result of that review.

Third, the company is required to create an audit committee that will create 'formal and transparent arrangements' for the manner in which financial reports and the results of internal controls will be put to work. As mentioned above, this committee should comprise three non-executive directors, one of whom should have 'recent financial experience' (Combined Code, C.3.1). The role of the audit committee is important. Its remit includes monitoring the integrity of the company's financial statements – at the heart of disclosure in securities regulation – as well as any announcements relating to the company's financial performance. The audit committee must also oversee the company's internal controls, its internal audit functions and its external audit process. It is the audit committee that should also review arrangements for whistle-blowers to raise concerns about financial misfeasance. The role of the audit committee is, then, something that should be explained in the annual report. In many senses, the work of the audit committee is one of the most important assets of corporate governance in that it seeks to ensure that all information relating to the company's finances that is made available to the public is accurate and reliable. Of course, in relation to Enron and WorldCom it was the failure of audit controls that contributed to the ultimate collapses of those organisations.

Relations with shareholders

The Code highlights the importance of good relations between management and shareholders. Consequently, it is provided that one of the responsibilities of the board of directors is to ensure that there is a 'satisfactory dialogue with shareholders' (Combined Code, D.1). Somewhat optimistically, perhaps, institutional shareholders for their part are expected to enter into dialogue with companies 'based on the mutual understanding of objectives' (Combined Code, E.1) as opposed to investing solely for short-term gain.

The effect of these principles on the culture of a company

The intention of the Code overall is to create a structure for the internal operation of the company which will minimise the risk of the company being mismanaged and which will also improve the way in which the company makes its senior management decisions. Bound up in all of this is an implied criticism of the ability of traditional company law to achieve effective corporate governance. In effect, what is being said is that the articles of association will not be enough to prevent abuses by corporate executives (in terms of their pay or falsifying accounts or mismanaging the company), and that instead what is important is *cultural change* within the organisation. And cultural change is said to come from proper corporate governance systems that cause executives to reflect more deeply on their decisions and to act with complete propriety because there are non-executives participating in key stages of the process. However, those non-executives also have a positive role in that they are frequently called on to advise the company how to proceed, and so they do not exist solely as guardians of the company's welfare. This is why so many non-executive directors of public companies are drawn from the executive boards of other public companies; the chairman of large public companies is usually a former CEO of a public company who is well placed to advise the CEO.

It might be thought that these two objectives are potentially in conflict: the urge to advise and the urge to police. In a perfect world, these two activities would be complementary: one can advise someone how to make a profit in a way that is in line with the proper running of the company. However, when it goes wrong, it will simply require either a larger conspiracy or a group of like-minded executive and non-executive directors all making the same decisions together. After all, if the directors all come from similar backgrounds and have similar careers, then it is likely that they will agree on questions like the appropriate size of a director's remuneration and what constitutes ethical behaviour. It all depends on the culture of the organisation: whether this corporate governance system promotes genuine scrutiny of decisions and genuine questioning of executive actions, or whether one group of directors simply rubber-stamps the decisions of the other group. Culture is a difficult territory for lawyers. Lawyers typically prefer rules to soft law codes because rules are easier to interpret and to apply. However, a large amount of corporate law practice involves an understanding of the culture of organisations and an understanding of the way in which high-level principles should be interpreted and acted out in given situations. In an uncertain world where change is the only thing we can be sure will happen in the future, understanding how to cope with flexibility is vitally important.

The lack of legal effect of corporate governance codes

If corporate governance standards of this sort are thought to be so important, perhaps they ought to be codified in legislation. At present, requirements set out in corporate governance standards will not even found any liabilities at common law, as Parker J held in *Re Astec (BSR) plc* (1998) when dismissing an action brought by a minority shareholder claiming unfair prejudice. Even though Parker J recognised that 'members of the public buying shares in listed companies may well expect that all relevant rules and codes of best practice will be complied with in relation to the company', his Lordship held that the code cannot found a petition for unfair prejudice.

Thinking about corporate governance

An alternative to the Anglo-American model

Twentieth-century corporate theory was generally divided between two different models of the company: the German–Japanese model and the UK–US model. The corporate model that has been used traditionally in Germany and Japan is one in which control is split between a management board made up of executive directors and a supervisory board which has involvement from employees, trade unions and others. Consequently people with a stake in the company but who were outside the executive board of directors were able to have a direct influence on the direction of the company. This suggests a more co-operative culture within the company between management and labour as a result. Then again, the commercial culture in Germany tends to rely on supportive structures more than the culture in the UK does. For example, banks lending money to companies tended to take a long-term view of their role with that company and to be more supportive of it.

In this book we have focused on what would be described as the UK–US model in this theory (although it is important to note that because of the advent of EU regulation of companies, there is now not the difference between the law in Germany and the law in the UK that there used to be). This second model is the one identified by Berle and Means (1932), in which the *ownership* of the company through the holding of shares is diversified among many shareholders but yet the *control* of the company is concentrated in the hands of a few managers. Some commentators doubt that this is the case in all circumstances, although it does appear to be a reasonable definition of the structure of public companies in the UK in that the business is run entirely by the directors (and other employees), with the shareholders tending to be investment funds and private investors. On this model there is only one board of directors (unlike the German two-tier board) and consequently day to day control of the company rests with the directors.

In essence, progressive UK and US theorists would point to the German model as being more progressive than that used in the UK and the USA. The UK–US model seeks to maintain free markets and to give boards of directors (and their shareholders to a lesser extent) greater freedom. By contrast, the German model recognises the interests of other stakeholders in the fortunes and the behaviour of the business. This is something that we shall consider to be in tune with the movement towards greater corporate social responsibility in Chapter 14.

Competing priorities within the company

One thing that corporate governance will not do is to remove conflict between different cliques within a company. There may be different agendas between shareholders on the one hand and directors on the other, as discussed in Chapter 6, for example if the shareholders want a quick dividend while the directors want to invest any spare cash in the business (Roe 2000). This phenomenon arose many times in January 2011 when directors had been careful to cut costs in their businesses during the recession, with the result that the business had excess cash at the end of the year. Many of the shareholders wanted this cash to be paid out to them in large dividends, although the directors typically wanted to retain that cash to guard against the future effects of the recession or to invest in the business.

If directors are in control of the business, they are likely to be concerned with the long-term health of that business (because it employs them) and perhaps will want to use spare cash to increase the salaries of the employees. By contrast, the shareholders may be interested in maximising short-term profits so as to increase their dividends or the market value of their shares. In contrast, it is thought that dividing control between a management board and a supervisory board (as in the German model) will create a balance between long- and short-term goals for a company because all of the participants in the company have a voice on the board. Involving employees in management decisions is thought to reduce industrial relations problems and also to show the company as being a place of employment as well as a source of profit.

The effect of deregulation and a lack of scrutiny

The introduction of corporate governance standards took place at the same time as financial markets and other business sectors had seen a reduction in regulation. The result could be understood as a muddy middle ground for the regulation of the whole of commercial life somewhere between formal law and no regulation at all. There was merely 'light-touch' regulation of corporate activity that was built on a policy of reducing governmental interference and regulatory oversight. This movement has been criticised by Professor Joseph Stiglitz in *The Roaring Nineties* (2004) and Professor Paul Krugman in *The Great Unravelling* (2003). Indeed many blamed the global financial crisis of 2007–10 on this process of deregulation. Corporate governance standards clearly represent a compromise between having no objective governance of companies at all and having a full-blown statutory code.

It was the Enron collapse which really turned this topic on its head, particularly in the USA. Whereas previously there had always been an undercurrent of this debate to the effect that management was essentially honest, even if it might be considered to be slightly self-interested, when it emerged that senior management at Enron had in fact been falsifying accounting records on such an enormous scale, it was no longer possible to be quite so sanguine about the intentions and probity of the management of large public companies. After all, if the policy was to increase the number of ordinary citizens who would invest in such companies, it was essential that the information being published for those markets was reliable. Consequently, it became obvious that greater oversight of the audit function, the remuneration of directors, and so on, would be required. Indeed it is difficult to discuss this topic without reflecting on the recent global financial crisis. There had previously been a consensus among economists for the most part that free markets would ensure that everyone behaved honestly and that the most worthy businesses would survive. However, even the much-lauded Alan Greenspan of the Federal Reserve Bank of New York acknowledged before a congressional committee that this market could not be relied upon to ensure the integrity of free markets on the basis of enlightened self-interest after all.

It has been argued by others that the Enron debacle was proof that what was needed was even freer markets and even less regulation, so that such bad practices could be worked out naturally by those markets (see Bratton 2002). In light of the recognition even by people like Mr Greenspan that free markets cannot be relied upon for a sort of informal regulation of commercial activity, that suggestion seems insupportable. The freedom of management at Enron allowed that collapse to happen.

Has the Anglo-American model really won?

There has been a suggestion by US academics Hansmann and Kraakmann that we are approaching 'the end of history for corporate law' (Hansmann and Kraakmann 2000). Their suggestion is that corporate governance thinking is converging on a single point. Moving on from Coase's theory that the company is a nexus of contracts, they suggest that all corporations should be run entirely for the benefit of their shareholders. In consequence, the directors' principal goal is to increase shareholder value. This was a feature of management theory throughout the 1990s, especially by theorists like Milton Friedman (1962), who argued that there is no purpose for a company other than to generate profits and to share those profits out among the shareholders. Consequently, corporate governance on such a world-view would be limited to the directors' obligation to earn profits at the expense of any other consideration.

As is discussed in Chapter 14, it is more common today to think about the impact that companies are having on communities and on the environment. Free-market economists such as Milton Friedman argued that worrying about corporate social responsibility was not something that management should waste its time on precisely because their sole goal should be to generate profit, and not to hand those profits out to good causes. Others such as Robert Reich (2008) have argued that when companies purport to be doing good works of this sort, they are in fact just carrying out specious public relations exercises that *are* ultimately directed at making greater profits.

Nevertheless, for Hansmann and Kraakmann, the argument runs that it is the US model of the corporation that has won out around the world. Consequently, they say that we have moved beyond worrying about labour-orientated or management-orientated approaches to the best form of corporate governance, and instead the consensus in corporate law models is that management should be focused solely on earning profits to be distributed among the shareholders. This is why they talk of it being 'the end of history' for corporate law, because a consensus has been reached in which the sole purpose of directors within a corporation is to earn profits.

This talk of the end of history is itself borrowed from a book by Francis Fukuyama called *The End of History and the Last Man* (1993). The over-hyped thesis of that book was that after the end of the Cold War, the American model of society and of capitalism had won, with the result that there would be no more history. The events of 9/11 and the ensuing wars have demonstrated that history is far from over. Therefore, it is somewhat peculiar that exactly the same assumption is being made about corporate law theory at a time when, as is illustrated in Chapter 14 of this book, there are competing models of what companies exist to do. Indeed, the financial crisis of 2007–10 suggests that the American model of capitalism is fatally flawed: particularly the model of capitalism popular in the 1990s, which fuelled a short-term boom and the ensuing collapse in which governments in developed capitalist countries were required to nationalise a number of their banks (Hudson 2009b, Chapter 32).

There is more in corporate theory than is recognised by your philosophy

There is in fact greater differentiation between corporate models around the world than is often acknowledged in this sort of scholarship. For example, in the UK and in Germany there are many different types of board structure. Indeed, Japanese company law has tended to adopt more American models. This is the convergence to which the American academics have been referring. All that this tells us, however, is that there is more in heaven and earth than is known in our philosophies.

One of the principal difficulties with the study of corporate governance, it is suggested, is that the subject matter itself is soft law and that the sources on which so many of the academics rely are so insubstantial. Many of these theories are built on vague economic histories of the world and on concepts like economic geography. In truth, what is being considered here is nothing more or less than the exercise of power within companies by management and the exercise of power through companies by capitalists.

Corporate governance scholarship is focused primarily on large public companies whose securities are quoted on stock exchanges and similar public markets. Yet, this tells us little about the more numerous types of small trading companies and asset management companies in existence. The corporate governance standards discussed in this chapter are of course applicable only to large companies, in which, in any event, one might expect some commonality of management approach. Corporate governance regimes are in truth hazy compromises between legislating to control the activities of boardrooms in large public companies (with all of the difficulties of political interference that that would seem to present) and simply allowing the possibility of directors indulging in practices that are considered by the public to be unacceptable.

Disqualification of directors

The disqualification of directors

One of the principal mechanisms for dealing with directors who act inappropriately is by disqualifying them from holding that office. In general terms, such a disqualification order has the effect that the defendant is prohibited from acting as a director during the period of the disqualification order, further to s 1 of the Company Directors Disqualification Act 1986 (CDDA 1986).

However, it is not simply that the defendant loses her parking space for a while. More specifically, the effect of a disqualification order is that the person in question may not, without the leave of the court, do any one of a number of things: she may not be a director of a company; she may not act as the receiver of a company's property; she shall not 'in any way, whether directly or indirectly, be concerned or take part in the promotion, formation or management of a company'; and she shall not act as an insolvency practitioner (CDDA 1986, s 1(1)). The maximum length of a disqualification order in any given set of circumstances depends on the context and the particular provision under which the order is sought, as discussed below, or else the court may accept an undertaking from a person not to act as a director or a receiver of a company's property for up to 15 years.

The four bases for disqualification in ordinary circumstances

Broadly speaking, there are four bases on which disqualification may be ordered when the company is solvent; each is considered in turn. It may be useful to understand the approach of the courts to these sorts of cases; in *Re Civica Investments Ltd* ((1983)) it was suggested that the approach taken in making a disqualification order was similar to sentencing in a criminal case.

Section 2 of the CDDA 1986 provides that when a director has been convicted of an indictable offence in connection with the promotion, formation, management, liquidation or striking-off of a company, or with the receivership of a company's property, or with her being the administrative receiver of a company, that person may be disqualified for a period of up to 15 years.

Section 3 of the CDDA 1986 provides that a person may be disqualified for up to five years from being a director of, or being concerned in the management of, a company if she has been 'persistently in default in relation to any provision of the companies legislation requiring any return, account or other document to be filed with, delivered, or sent to the registrar of companies' (CCDA 1986, s 3(1)). A person is taken to have been persistently in default in relation to any provision of the Companies Acts if she has been guilty of three or more defaults in the five years leading up to the application. For example, in *Re Arctic Engineering Ltd* (1986) failure to send 35 required returns to the registrar was held to be sufficient evidence for the making of an order.

Section 4 of the CDDA 1986 provides that if during a winding up it 'appears' that a person has been guilty of fraudulent trading or that she has been guilty of fraud or breach of duty in relation to the company while an officer of that company, that person may be disqualified for up to 15 years.

Section 5 of the CDDA 1986 provides that if a person is convicted of any offence as a result of failing to comply with any provision of the Act and has committed three defaults in the five-year period leading up to her conviction, she may be disqualified for a period of up to five years.

Disqualification orders following corporate insolvency

Now we turn to consider the disqualification of directors where the company is insolvent. (Insolvency in general terms is considered in Chapter 13.) Section 6 of the CDDA 1986 deals with disqualification of directors of insolvent companies on grounds of unfitness. Ordinarily, in such circumstances an order for disqualification must be for at least two years. The core question that the court has to consider is whether a defendant director's conduct, viewed cumulatively (that is, taking all behaviour into account) and taking into account any extenuating factors, fell below the standards of probity and competence appropriate for directors of companies. For example, in *Kappler v Secretary of State* (2008) it was held that there was sufficient evidence of dishonesty when it could be demonstrated that the director had known that false invoices had been raised by the company. Similarly, it was held in *Secretary of State for Trade and Industry v Thornbury* (2008) that directors cannot absolve themselves of responsibility if they leave all the company's financial matters to other people; that will not absolve them from the charge that they have acted without sufficient probity when those other people act dishonestly in relation to the company's finances. It was held that the director's high degree of incompetence, even if he had not himself been dishonest, justified a period of disqualification as a director. However, it was held in *Green v Walkling* (2008) that a director will not be found to have acted inappropriately or dishonestly if he acted throughout on the advice of his solicitor.

The application for such an order must be made by the Secretary of State, or in practice by civil servants in the appropriate department acting on behalf of the Secretary of State. The application will be made if it appears to the Secretary of State to be in the public interest as a result of information received from a liquidator, administrator or administrative receiver. The proceedings are civil, involving the civil standard of proof, and not the criminal standard (*Re Living Images Ltd* (1996)). Nevertheless they do involve penal consequences for the director, with the result that Art 6(1) of the European Convention on Human Rights applies as to the director's right to a fair trial within reasonable time (*DC v United Kingdom* (2000)).

Disqualification undertakings

The sheer volume of proceedings for disqualification orders following an insolvency put great pressure on the system. Therefore, a means was sought to speed up the process of disqualifying directors, known as the *Carecraft* procedure (first used in *Re Carecraft Construction Co Ltd* (1993)). It was adopted by the common law for non-contentious cases. This meant that directors could be disqualified without the need for lengthy court proceedings. Amendments were made to the CDDA 1986 (in ss 1(1A) and 7(2A)) to permit the Secretary of State to accept a disqualification undertaking from the defendant that she will not act as a director, receiver, promoter or manager of a company or as an insolvency practitioner for a stated period of between 2 and 15 years. In *Re Blackspur Group plc (No. 3)* (2002), the Court of Appeal held that in delegating this matter to the Secretary of State, Parliament had taken the view that the Secretary of State would form a perfectly acceptable appreciation of what was expedient in the public interest, such that the Secretary of State was quite entitled to refuse to accept an undertaking that did not have a schedule of the grounds of unfitness attached.

Unfitness to be concerned in the management

The grounds for deciding whether or not a person's conduct makes her unfit to be concerned in the management of a company are governed by s 9 of the CDDA 1986, which requires the court to have regard to the matters specified in Sched 1 to that Act. So the court must consider whether there has been: first, any misfeasance, breach of duty or misapplication of assets; second, failure to comply with legislative requirements as to the maintenance of books and records, returns and accounts; third, where the company is insolvent, responsibility for the cause of insolvency or for losses of customers who furnished advance payments and involvement in any transaction (or preference) which can be set aside; and, fourth, failure to comply with the statutory requirements relating to insolvency.

In *Re Dawson Print Group Ltd* (1987), 324, Hoffmann J held:

> There must, I think, be something about the case, some conduct which if not dishonest is at any rate in breach of standards of commercial morality, or some really gross incompetence which persuades the court that it would be a danger to the public if he were allowed to continue to be involved in the management of companies, before a disqualification order is made.

As Browne-Wilkinson VC put it in *Re McNulty's Interchange Ltd* (1988):

> Ordinary commercial misjudgment is in itself not sufficient to justify disqualification. In the normal case, the conduct complained of must display a lack of commercial probity although I have no doubt that in an extreme case of gross negligence or total incompetence disqualification could be appropriate.

In spite of these pithy summaries of the guiding principle, the Court of Appeal in *Re Sevenoaks Stationers (Retail) Ltd* (1991) held that they should not be used to replace the statutory wording. It was held, nevertheless, by that court in putting a gloss on the statutory wording that the incompetence involved was not required to be total incompetence, provided that it was more than a simple commercial misjudgment. It should involve some lack of probity or some abuse of the system. As Lewison J put it in *Secretary of State v Goldberg* (2004), the court is involved in considering the 'cumulative effect' of the allegations made against the director.

It is not enough for the defendant to cease acting in such an inappropriate manner. As Lord Woolf MR held in *Secretary of State for Trade and Industry v Griffiths* (1998), it was not enough to resist the making of an order that the defendant has shown that he has mended his ways and that he is no longer a danger to the public.

Period of disqualification

If the court finds that a director is unfit to be concerned in the management of a company under s 6 of the Company Directors Disqualification Act 1986, it has no choice but to impose a period of disqualification of between 2 and 15 years (CDDA 1986, s 6(4)). This is the only area where a disqualification order must be imposed; the court has no discretion except as to the length of the order. This was criticised by Vinelott J in *Re Pamstock Ltd* (1994) as being unduly rigid, especially as no minimum period is required in other areas such as fraudulent trading. There is no equivalent of a conditional discharge.

The Court of Appeal in *Re Sevenoaks Stationers (Retail) Ltd* (1991) laid down what, in effect, amounts to sentencing guidelines which divide between disqualification of over 10 years for very serious cases; disqualification for 6 to 10 years where there has been a serious breach of duty involving a misappropriation of assets or an abuse of the corporate form or something similar; and disqualification for less than 6 years for less serious cases such as gross negligence or incompetence. (See also *Re Vintage Hallmark plc* (2007).) It has been held, entirely correctly in this writer's opinion, that the length of the period is a matter for the court and not for agreement between the parties (*Re Barings plc* (1998)). It is not enough for the parties to 'privatise' the dispute and to decide the appropriate punishment between themselves: instead, this is a matter for the formal imposition of a penal sanction by the court.

Conclusion

In consequence, we can see that there are two ways in which corporate governance can be carried out: the effective way and the ineffective way. The ineffective way is a system of soft law like the Combined Code on Corporate Governance, which was created by consultation with the public companies and which those companies may choose to ignore if they choose to do so. This means in practice that public companies will ignore the code when economic conditions make it desirable to ignore it. By contrast, the rules on the disqualification of directors suggest a very different way of governing the behaviour of directors using positive law. In another dimension, the power to dismiss directors offers shareholders an ability to use a somewhat blunt instrument to deal with directors who do not meet their expectations.

Chapter 10

The company's capital

Introduction

One of the most complicated aspects of the organisation of a company can be the organisation of its capital. This is particularly true if there are many different types of shareholder or a number of competing interests inside the company. From the perspective of company law, however, it is also important to ensure that outsiders are not disadvantaged by the company's particular capital structure, and particularly to ensure that reorganisations of the company's capital structure do not leave a mere corporate shell behind which cannot pay third parties what they are owed by the company. This chapter considers the main legal principles relating to a company's capital, particularly its share capital, beginning with a brief introduction to the vagaries of corporate capital structures.

There are two central principles considered in this chapter. First, that a company's share capital must be raised and, once raised, must be maintained for the benefit of creditors. Second, the law then controls the circumstances in which there can be a reduction of a company's share capital. The current trend, however, is to regard the two central principles as being antiquated, and that they should be replaced by solvency and liquidity criteria. In essence, the issue is this. In the 19th century in particular, there was a high degree of fraud conducted in relation to companies. Therefore, the focus of the early case law was on ensuring that the statement of the company's capital reflected the company's actual capital base, in part by preventing the company from reducing its capital and in part by ensuring that the company's capital was appropriately paid up.

However, there is a modern view that company law should focus instead on the solvency of the company. Rather than focus solely on a company's net worth, finance theory prefers to analyse a company's cash flow (because this is less easy to manipulate). Furthermore, public companies like to be able to manipulate their capital base so as to enhance their capital position and to engineer earnings growth. Therefore, the Companies Act 2006 (CA 2006) contains a number of exceptions to the long-standing principles against manipulating a company's capital to give those companies a little more freedom.

The Company's Capital

The meaning of 'capital'

The central portion of a company's capital when that company is first created is the share capital that the original subscribing members of the company (that is, its very first shareholders) contribute to the company at the outset. Over time, the company's borrowings and profits which are held in reserve will also be added to the capital base of the company (the equivalent of moving money from your current account to your deposit account), and also any capital assets like buildings, business premises and plant and machinery which are acquired over time.

However, there are other, more complex accounting concepts that affect the capital and the income position of the company: for example, the need to account for the reduction in the value of capital assets over time ('depreciation') and the way in which cash or other assets are moved into reserves held for different purposes. In these accounting senses, the word 'capital' used in connection with a company has several different meanings; thus it may mean the nominal or authorised share capital, the issued or allotted share capital, the paid-up share capital or the reserve share capital of the company. Each of these terms relates to different aspects of the company's capital. We shall consider the meaning of each in turn.

The 'nominal' or 'authorised' capital is merely the amount of share capital that the company is authorised to issue, but that is not necessarily the same as what it has actually issued. So, a company may choose to grant the directors the power to issue 100,000 shares with a nominal value of £10 each, but may actually decide to issue only 50,000 shares at the outset because that is all the capital that is needed at that time or because only a limited number of investors can be found. In the case of a limited company the amount of potential share capital with which it proposes to be registered, and the division of that capital into individual shares of a fixed value (the nominal amount) must be set out in the company's constitution, but this amount may be increased or reduced, as explained later. Companies may fix on any figure that is large enough for its potential requirements. Public companies must have at least £50,000 in authorised capital (s 761(2) CA 2006).

By contrast with the nominal share capital, the 'issued' or 'allotted' capital refers to that part of the company's nominal capital that has actually been issued or allotted to the shareholders. A private company is not bound to allot all its capital at once, whereas a public company must allot at least £50,000 by way of the nominal value of shares before it may commence business (s 761 CA 2006). (The law concerning the allotment of shares is considered in Chapter 5.) Allotments of capital are made as they are needed by the directors, provided that the directors have been properly authorised to do so under the Act, up to the amount identified in the company's constitution as that company's authorised capital. There is a subtle difference between a share being allotted, which is when the company and the shareholder have agreed contractually to issue and acquire those shares, and it being 'issued', which is when the shareholder's ownership of that share is registered by the company. There are circumstances in which that distinction may be important. For example, the restrictions and controls imposed by the Companies Act 2006 relate in general to the allotted shares. In the context of taxation the distinction may also be significant, as in *National Westminster Bank plc v I.R.C.* (1994), in which a majority of the House of Lords held that a share is allotted when the company and shareholder are contractually bound but it is only issued when the shareholder is registered as such on the register of members, the liability to tax falling on one but not the other.

The 'paid-up' capital is that part of the issued capital that has actually been paid up by the shareholders. There may clearly be a time between a shareholder agreeing to acquire shares and the payment for those shares being made. There are controls on the company seeking to issue shares at a discount or issuing shares with payment only needing to be made later: otherwise, clearly, the company could claim to have a much larger share capital than it had actually received in cash. As a company grows, it may seek to issue blocs of shares as it needs to draw in more capital. Having the power to issue shares in advance makes it easier for the directors to issue shares as and when they need to do so. Suppose a company has a nominal capital of £500,000, which would be divided into 500,000 shares of a nominal amount of £1 each. The directors may decide that the company only needs £400,000 and therefore only 400,000 shares would be issued, with 100,000 shares that the directors have the power to issue in the future being held back. Alternatively, if the directors only required £100,000 immediately, or if their shareholders agreed with the company that they would only pay for their shares in bits, then 400,000 shares may be issued, but only £100,000 actually be paid up in that the company has so far required only 25p to be paid up on each share. Therefore, the issued share capital would be £400,000 but the actual paid-up share capital at that time would only be £100,000. The uncalled (that is, not paid up) capital is the remainder of the issued capital and can be called up at any time by the company from the shareholders in accordance with the provisions of the company's articles. Section 586 of the CA 2006 requires public companies to call up at least one quarter of the nominal value of a share and the entire premium on allotment.

It is possible that shares will be sold for an excess over their nominal value. The paid-up share capital includes the nominal value of the shares paid up and any premium on such shares. For example, if 10,000 £1 shares are sold for £2 each, the paid-up capital will be £20,000. The excess £1 paid for each share is then paid into a 'share premium account' distinct from the nominal share capital. So, this will be expressed as £10,000 share capital and £10,000 share premium account in the balance sheet. It has even been held that where the shareholders agree to increase a company's capital without any formal allocation of shares – so that each shareholder contributes more money to the company's share capital without receiving more shares in return – that increase will be treated like a share premium and so be subject to the capital maintenance rules (*Kellar v Williams* (2002)).

Shares must not be issued at a discount

As part of maintaining a company's capital, it is required that the company receives the full subscription price of the shares from its shareholders. The general rule has always been that shares must not be issued at a discount. Thus, a share may not be issued as being 'fully paid' if in fact the amount paid for that share is less than the nominal amount set out in the company's constitution. The general principle was first established in England by the House of Lords in the 19th-century case of *Ooregum Gold Mining Co. of India Ltd v Roper* (1892). In that case the market value of the £1 ordinary shares of a company was 2s 6d (that is, two shillings and sixpence). The company purported to issue preference shares of £1 each with 15s credited as having been paid (thus leaving a liability of 5s per share to be paid to make up the £1). Therefore, the shareholders would have been getting their preference shares at a discount to the price specified in the company's articles of association. It was held that the issue was beyond the powers of the company and consequently that the purported shareholders were required to pay for the shares in full. As Lord Macnaghten held:

> The dominant and cardinal principle of [the Companies] Acts is that the investor shall purchase immunity from liability beyond a certain limit on the terms that there shall be and remain a liability up to that limit . . . It is plain that this [principle] is one which cannot be dispensed with by anything in the articles of association, or by any resolution of the company, or by any contract between the company and outsiders who have been invited to become members of the company and who do come in on the faith of such a contract [at p 145].

This principle is now expressly set out in s 580 of the CA 2006. The shareholder must pay the full nominal value of his shares, either in cash or in kind. If the share is not paid for in full at the outset, the shareholder owes the balance plus interest to the company. The court may grant relief in appropriate circumstances if it considers it 'just and equitable' to do so (s 589(2) CA 2006). Contravention of s 580 is a criminal offence by the company and any officer in default (s 590 CA 2006).

Controls on the company's participation in the acquisition of shares

Introduction

This section considers two important restrictions on dealings with a company's shares. In essence, the underlying purpose of both schemes of rules is to protect the capital base of the company. The first restriction precludes 'financial assistance', which occurs when a company provides value either directly or indirectly to enable someone else to become a shareholder in that company. If the company were to do this, it would mean that the company's capital was being reduced by transferring money or other assets so as to acquire shares, or simply by increasing the number of shares that are nominally in issue without increasing the amount of money that has been brought into the company. Effectively these would be 'free' shares from the shareholder's perspective, because they would not have paid for them in full with their own money. The second restriction precludes the company from acquiring its own shares. Another means of reducing the company's capital would be to refund the price of shares that had already been issued. There may be contexts in which the company might want to reduce the number of shares that are in issue, but that must be done in accordance with the statutory procedure discussed below.

If companies were able to reduce their capital in either of these ways then that would create the risk of fraud. Suppose that a new company was being launched with a view to raising capital from the public to fund a new business venture. To attract investors it would be advantageous if the company already had a large amount of investment capital pledged to it. But if the promoters of the company had arranged with these first investors that the company would buy their shares back in the future or that the company would lend those investors the money with which they could buy shares, that would mean that the investing public would be misled as to the amount of capital that was invested in the company, and the company's trade creditors would also be misled as to how much capital there was in the company in the event that the company might fail in the future. So English company law developed a policy from its earliest years (as discussed in Chapter 1) to ensure that companies could not tinker with their capital in a way that would be misleading to others.

Financial assistance

Financial assistance is the process by which a company makes a gift of money or a loan or otherwise enables another person to buy that company's shares. It is a practice with a long pedigree. The giving of financial assistance by a company to purchase that company's shares is prohibited by s 678 of the Companies Act 2006 (CA 2006). We shall first consider this statutory prohibition and then try to plot a course through the large amount of case law that has considered the concept of 'financial assistance' in the predecessor legislation.

The statutory prohibitions on financial assistance

There are three prohibitions on financial assistance in the Companies Act 2006. The first prohibition in s 678(1) of the CA 2006 provides that:

> Where a person is acquiring or proposing to acquire shares in a public company, it is not lawful for that company, or a company that is a subsidiary of that

company, to give financial assistance directly or indirectly for the purpose of the acquisition before or at the same time as the acquisition takes place.

Thus financial assistance in relation to public companies or their subsidiaries is unlawful. As will emerge below, there is a conditional exception in relation to assistance provided by private companies (s 682(1)(a) CA 2006): this was a significant change from the law before 2006. The term 'financial assistance' is defined in s 677 CA 2006 so as to include assistance by way of gift, or by way of guarantee or indemnity, or by way of a loan, or by way of any other arrangement such that the company's net assets are reduced (thus implying that value has shifted from the company to some other person so they are able to acquire shares).

The second prohibition on financial assistance arises under s 678(3) CA 2006 in circumstances in which the purchaser of shares takes on a liability in so doing which the company then undertakes to reduce or expunge. This would cause the company to pass value indirectly to the purchaser, which would constitute financial assistance. The third prohibition arises under s 679(1) CA 2006 if a person acquires shares in a private company and receives financial assistance from a subsidiary of that company. A further prohibition arises under s 679(3) where a liability is reduced to facilitate such an acquisition.

The CA 2006 has therefore become more permissive in relation to private company transactions than had been the case previously. Nevertheless, the interpretation that the courts have tended to give to the concept of financial assistance has been so broad in many cases that a number of transactions that would prima facie appear to fall outside its mischief have been caught within its net. The principles that have emerged from that case law are considered below, after the exclusions from the statute are considered in the next section.

Statutory exclusions

There are specific statutory exclusions from this prohibition on financial assistance: that is, there are situations in which financial assistance will be permitted. (There is something very English about drafting statutes using double negatives in this way: 'thou shalt not do this . . . except that thou need not not do this in the following situations'.) The first permission appears in s 678(2) CA 2006, which provides that the prohibition in s 678(1):

> does not prohibit a company from giving financial assistance for the acquisition of shares in it or its holding company if
>
> (a) the company's principal purpose in giving the assistance is not to give it for the purpose of any such acquisition, or
> (b) the giving of the assistance for that purpose is only an incidental part of some larger purpose of the company,
>
> and the assistance is given in good faith in the interests of the company.

Therefore, this exception is based on the assistance being something other than the 'principal purpose' of a transaction – which is a theme that emerges in the case law below. The leading case *Brady v Brady* (1989) took a narrow approach to collateral purposes of this sort, although the statute now permits the prohibition to be eluded if the assistance was not the principal purpose of the transaction.

Furthermore, s 681(2) CA 2006 sets out eight specific transactions permitted under company law that are not to be subject to the prohibition on financial assistance. These are as follows:

(1) the distribution of lawful dividends;
(2) distributions made in the course of a winding up;
(3) the allotment of bonus shares;
(4) any reduction of capital confirmed by the court;
(5) a redemption of shares or a purchase of its own shares by a company;
(6) anything done under a court order under Part 26 of the CA 2006;
(7) anything done under an arrangement between a company and its creditors under s 1 of the Insolvency Act 1986;
(8) anything done under an arrangement made in pursuance of s 110 of the Insolvency Act 1986.

There are also two species of exception under s 682 CA 2006: first, if the company is a private company (s 682(1)(a)); second, in relation to assistance given by public companies where their net assets are not reduced or where any reduction in those assets is paid out of distributable profits (s 682(1)(b)).

The case law on financial assistance

What emerges from the case law is that the financial assistance must be financial in nature, as opposed to being some other sort of assistance (*Barclays Bank v British and Commonwealth Holdings plc* (1996)), and that it must be given for the purpose of acquiring shares (*Brady v Brady* (1989)). The case law was decided before the CA 2006. *Brady v Brady* was part of a line of cases in which a strict approach was taken such that even inadvertent financial assistance fell within the net. Contrariwise, another line of cases, including *Charterhouse Investment Trust Ltd v Tempest Diesels Ltd* (1986) and *MT Realisations Ltd v Digital Equipment Co Ltd* (2002), took the approach that if the financial assistance was not the main purpose of a larger transaction, the prohibition on financial assistance ought not to be used to frustrate that larger transaction. The latter line of cases appear to resemble the statute more closely than the stricter line of cases now.

So, in *Brady v Brady* (1989), to make a long story short, two brothers fell out with one another, which raised problems for the organisation of their family's businesses. There were two private companies that ran the family's businesses and therefore a decision was taken to give the brothers one company each. However, one company had assets that were considerably more valuable than the other. To achieve equality between the brothers, it was agreed that one company would transfer sufficient assets to the other company by means of paying off loan stock that had been issued as part of the purchase price for shares. However, because these assets had been used to pay off borrowing which had been taken out to buy shares in the first place, this appeared to be financial assistance which had made the acquisition of shares possible. Clearly, this was incidental to the main purpose of the transaction, which was to achieve parity between the brothers in the division of the family business. It had been accepted by the Court of Appeal that any financial assistance was incidental to the parties' principal objectives. Nevertheless, the House of Lords held that there had been financial assistance on these facts and that the broader context of the transaction did not affect that analysis. Lord Oliver held that even though the *reason* for the transaction was to reorganise the group of companies, nevertheless the *purpose* of the redemption of the loan stock was simply to acquire shares, and therefore it constituted financial assistance.

The contrary approach emerged from cases like *Charterhouse Investment Trust Ltd v Tempest Diesels Ltd* (1986), in which a company, as part of a management buy-out, had sold all of the shares in a subsidiary company to one of its directors and thus had sold a part of its business to that director. This was found by Hoffmann J to have been effectively the principal purpose of the transaction. The issue was whether a transfer of tax losses between the two companies when the subsidiary was hived off would constitute financial assistance. Tax losses are valuable assets because they reduce the amount of tax that their owner has to pay. (Sometimes failing companies are taken over simply because their tax losses will save the purchaser so much in tax.) Hoffmann J held that the transfer of the tax losses in this case constituted a mere part of a much larger transaction and therefore could not be viewed as having been financial assistance in isolation. Hoffmann J suggested that the acid test for financial assistance should be whether or not the company's net assets were reduced by the transaction. Hoffmann J held that the question, in essence, was whether there had been a net transfer of value by the company to the acquirer. Because such a reduction in the net worth of the company could not be shown on those facts, it was held that there had been no financial assistance.

Another case that sought to limit the legislation on financial assistance (before the CA 2006) was a decision of Laddie J. So, in *MT Realisations Ltd v Digital Equipment Co Ltd* (2002), Laddie J held that there should be a distinction drawn between acceptable behaviour such as a company giving financial incentives to someone as part of a larger transaction or doing that so as to induce someone to enter into a transaction, and unacceptable financial assistance for the actual acquisition of the shares. This approach was, however, roundly criticised by the Court of Appeal in *Chaston v SWP Group Ltd* (2003), where a company had paid the fees for the drawing-up of a report on its parent company for the benefit of a prospective purchaser of the parent company's shares: it was held that paying for that report had in fact been part of assisting the purchaser to make the purchase of those shares. Rejecting Laddie J's approach, the Court of Appeal held that this payment had assisted the purchaser, had helped to smooth the course of the acquisition and was intended to further to the main transaction. Consequently, it was held that it must have been made for the purpose of the acquisition.

Similarly to *Charterhouse*, in *Dyment v Boyden* (2004), the only two shareholders in a company that ran a nursing home agreed, when one of them had been disqualified from operating a nursing home, that the disqualified shareholder would transfer his shares to the other party but that the company would lease property from him at a rent which was higher than a market rate. The issue arose whether this payment was in truth financial assistance. It was held by Hart J that in these circumstances the parties were simply identifying how they could best run the nursing home business for the future and were not intending to provide one shareholder with financial assistance after the acquisition of the shares. Any financial assistance was therefore incidental to the principal purpose of the transaction. (This was affirmed by the Court of Appeal (2005)). In effect, both Hart J and the Court of Appeal held that the excess rent was not unlawful financial assistance for the transfer because, although it was connected with the sale of the shares; it was not for the purpose of that sale in a commercial sense. Instead, the principal purpose of the company had been to enter into an agreement that would allow its business to continue once one of the shareholders had lost his licence to trade. While that shareholder's personal objective may well have been to make up for his loss of income from the company, that was nevertheless not the company's objective.

It is worth noting how effectively the *Salomon* principle operated in the minds of the judges here. The company was operated by two human beings, one of whom they accepted wanted to replace his lost income by means of this transaction. Nevertheless the court managed to draw an artificial distinction between that human being's objectives and the objectives of the company: even though the intention in that human mind constituted at least half of the intention of the two people involved in the company.

By contrast to *Dyment* but applying the same approach in effect, in *Harlow v Loveday* (2004), it was held that when the parties to a loan knew that the money was intended to be used to buy shares in the company, an argument that there was no compulsion to use the loan moneys for that purpose would not blind the court to the fact that the loan was intended to be a form of financial assistance. In essence, the courts are examining the substance of the transaction and not allowing the case to be decided by whether or not the transaction involved something that could be taken literally to involve a financial transaction. So in *Harlow v Loveday*, it was held that the substance of the transaction was actually financial assistance.

A good example of the sorts of considerations which emerge in questions about whether or not the financial assistance was the principal purpose of a transaction is the case of *Belmont Finance Corporation v Williams Furniture Ltd (No 2)* (1980), where Buckley LJ was called upon to consider a situation in which a company had bought goods from a supplier and the supplier had then used that money to buy shares in the company. His Lordship considered that this should not fall within the prohibition on financial assistance because the principal purpose of the transaction had been the company's need for the goods which it had acquired, and therefore the payment of money to the supplier had constituted a part only of a larger transaction, and consequently it had not been intended solely to assist the purchase of shares. However, if the company had not required the goods and had acquired them merely as a means of paying money to the supplier so that the supplier would be able to acquire the company's shares, then there would have been financial assistance. Significantly, then, the company had genuinely required the goods and had not entered into the transaction as a ruse to conceal financial assistance.

Restrictions on the company acquiring its own shares

Capitalism has changed greatly since the Industrial Revolution, although there are many essential truths in that early law which remain valuable today. The decision of the House of Lords in *Trevor v Whitworth* (1887) established the principle that the company should not be permitted to buy back its own shares from shareholders because that would mean that the company's capital would be reduced to pay for the recovery of those shares. This was clearly a bulwark against fraudulent schemes to raise money from the public while repaying conspirators their initial investments in the company by a share buy-back scheme. This is an example of an essential truth that has stood the test of time: companies need to be prevented from having a free rein to do something that could facilitate fraud by unscrupulous rogues.

Now, in the 21st century, capitalism has become more orientated around sophisticated financial products to raise and manipulate capital for large public companies. Indeed, the financial crisis of 2007–10 was caused by just such financial instruments, which marketed large numbers of sub-prime mortgages as complex securities and shifted the risks associated with those mortgages off the balance sheets of the large banks, while idiotically putting those risks back on those same balance sheets through credit default swaps, which all but insured payments under those same mortgages. In this same spirit of increasing use of sophisticated financing techniques, it has become an attractive strategy for some companies with surplus cash to use that spare cash to buy back shares from shareholders. For example, during a recession companies will tend to cut their costs, which means that as the economy improves, there will be surplus cash. Companies will not want to pay an excessively large dividend in one year because it will look in subsequent years as though their dividends and thus their profitability had decreased. So management will look for ways of diverting that surplus cash into an investment in the business. Buying back shares has the effect of reducing the amount of the company's equity capital so that it is possible to raise more equity in the future if it is needed, but it provides a use for spare cash if there is no other suitable investment at the time (perhaps if the economy is sluggish). A company with a large amount of cash in its bank accounts becomes a target for takeovers. So one way of 'saving' that cash is to keep it on the balance sheet as capital by reducing the number of shares that are in issue.

You could think of it as being a distant cousin of the idea that a private individual would not put cash into a deposit account at a time when she has large credit card debts: she would use that spare cash to pay off her credit cards, which would leave her with more credit to use in the future and would avoid her paying large interest payments in the short term. Corporate finance is led now by complex business theory, accounting techniques and mathematics which seek the most effective ways of producing a balance sheet and cash flow analysis that paint a company in the most attractive light. Consequently, there has been a call for company law to relax many of its restrictions on the manipulation of corporate capital. In so doing, we should not forget all of the lessons of history – both distant and recent.

The statutory restriction on company acquiring own shares

The principle that a company cannot reduce its capital except in accordance with the Companies Act 2006 means that it is unlawful for a company to buy its own shares or to redeem its own shares without statutory authorisation, irrespective of the extent to which such a purchase is in the interests of the company or its shareholders as a whole (*Trevor v Whitworth* (1887)).

Further to s 658 CA 2006, a limited company may not acquire its own shares, 'whether by purchase, subscription or otherwise', unless one of the statutory exceptions to that rule applies. Four exceptions to this general rule are set out in s 659 CA 2006: first, a limited company may acquire any of its own fully paid shares 'otherwise than for valuable consideration'; second, in relation to a reduction of capital under s 135 CA 2006; third, in relation to purchases authorised by the court under ss 98, 721 or 759 CA 2006; fourth, in relation to the forfeiture of shares or acceptance of any shares surrendered in lieu of failure to pay for them. It has also been held at common law that the acquisition of a company that owns shares in the acquiring company falls outside the central principle (*Acatos & Hutcheson plc v Watson* (1995)).

Another complex area relates to 'redeemable shares' issued by a company: that is, shares that can be cancelled according to their own terms. Section 684(1) CA 2006 provides that a limited company 'may issue shares that are to be redeemed or are liable to be redeemed at the option of the company or the shareholder'. A public company may only do this if it is empowered to do so by its articles of association; and any articles of association may preclude the issue of redeemable shares. One situation in which redeemable shares may be issued would be if a venture capitalist wanted to invest in a company by acquiring shares but also wanted to be able to remove its investment on the happening of an identified event (such as the company's net worth falling to a given amount, or the venture capitalist simply deciding it wanted to place its investment elsewhere). That shares can be redeemed means that the shareholders involved can extricate themselves from the company without having to find a buyer for their shares. There may be conditions placed on the redemption of the shares in the company's articles of association to prevent the company's share capital being reduced randomly.

Further to s 690 CA 2006, a company may purchase its own shares provided that it does so in accordance the statutory procedure and provided that there is no restriction on the company doing so contained in its articles of association. There are three statutory limitations on the right to repurchase shares: the shares must be fully paid up; the effect of the repurchase must not be to leave no paid-up or treasury shares in issue; and the shares must be paid for on purchase out of distributable profits or by the issue of fresh shares (s 690(2) CA 2006). Otherwise the parties may reach whatever terms for the repurchase they wish.

Treasury shares

Treasury shares are shares that are held by the company in treasury, a sort of abeyance, pending disposal or cancellation. The ability to diminish or increase the number of its shares owned by shareholders is seen as an advantage to companies in managing their capital structures, in particular their debt/equity ratios, which are usually significant in their corporate finance covenants. Changes made in 2003 allowed some companies the option to retain and to reissue the shares that they had purchased. These regulations were re-enacted without substantive change in ss 724 *et seq.* of the CA 2006.

Open-ended investment companies

The principal exception to the notion that a company may not buy its own shares is the open-ended investment company (the 'oeic'). Oeics were introduced to English company law so corporate equivalents to unit trusts could be marketed from the UK in jurisdictions where there are no trusts. (For a discussion of oeics, see Alastair Hudson, *The Law of Finance* (Sweet & Maxwell, 2009), paras 52–28 *et seq.*) Investors acquire shares in an oeic, and the operators of the oeic then invest the company's assets and pay the profits out among the shareholders. For an oeic to function, it is important that its investors are able to terminate their investments by selling their shares back to the company when they choose individually that they wish to do so. Thus, an oeic must be able to buy back its own shares whenever an investor requires it too. Consequently, oeics are exceptions to the rule against companies buying back their own shares.

Debentures and company charges

The nature of debentures

There is always a question as to whether or not a company has the power to borrow money. In the ordinary course of events, as discussed in Chapter 4, no transaction will be void on the basis of the capacity of a company provided that the lender is acting in good faith (thus bringing the transaction outside s 40 of the 2006 CA 2006) and that the lender is not a director of the company (within s 41 of the 2006 CA 2006).

Traditionally, companies have often borrowed money by means of debentures. A debenture is basically a loan which grants particular rights to the lender to receive periodical payments of interest and the repayment of the loan capital on the termination of the transaction, as well as further security rights which are embodied in documentary form. This usually means that that the company creates either a fixed or a floating charge over its property as security for the loan. Debentures may be effected by issuing 'debenture stock' to the lenders, which is very similar to issuing bonds in the manner discussed in Chapter 11. When debenture stock is issued, there will be a large number of investors acquiring that stock in units in proportion to the size of their investment, as with a bond. For present purposes there is little practical difference between a debenture and a bond, except that debentures have a longer history in English law. Today, however, it is the bond markets that are important in relation to large public companies, as discussed in Chapter 11.

The problem of defining a debenture

A debenture is defined in *Charlesworth's Company Law* as being 'a document which creates or acknowledges a debt due from a company' (paragraph 25–003). Classically, Chitty J held that the term 'debenture' could not be defined: 'I cannot find any precise definition of the term, it is not either in law or commerce a strictly technical term, or what is called a term of art' (*Levy v Abercorris Slate and Slab Company* (1883)). What his Lordship meant was that debentures could take so many forms that a definition in the abstract would be difficult to create. Ordinarily in practice a debenture would be in the form of a document under seal (that is, a document to which the company's seal has been fixed in accordance with s 45 CA 2006).

The role of the debenture trustee

In many circumstances a debenture will also be supported by a trust such that a trustee is appointed to ensure the proper performance of all obligations under the debenture. The trust is structured in the following way: the trustee takes legal title in the debenture stock and so holds the benefit of the debenture, and the benefit of any charge effected as part of the debenture arrangement, on trust for the investors. As with all trustees used in Eurobonds transactions, the role of the trustee is really a fiduciary with powers under the trust deed to enforce the borrower's obligations to pay interest. This is particularly useful where there are a number of investors who have acquired debenture stock from the company as part of a single issue of debenture stock, because the trustee is empowered to act on behalf of all of those investors to resolve any disputes with the borrower. A good example of this power in action arose in the recent cases of *Law Debenture Trust Corporation v Elektrim Finance BV* (2005) and in *Citibank NA v MBIA Assurance SA* (2006), where the trustees activated termination provisions under the loan documentation that required the borrowers to repay the loan capital immediately and would have activated any rights under a charge.

An investor who holds a debenture is both a lender to the company with rights under the law of contract and also a creditor of the company under the debenture with any rights created by the debenture itself. In the ordinary course of events, if the loan is repaid, then the rights under the debenture have no practical effect and will fall away when the loan capital and all interest has been paid by the borrowing company.

It is more common in financial parlance when considering public companies to focus on bonds that are issued by those companies, as discussed in Chapter 11. Borrowing by way of debentures still takes place, although more often in relation to smaller companies. It should be remembered that the *Salomon* litigation, for example, began with a creditor seeking repayment from Aron Solomon himself of loans that had been made to a company by way of a debenture. However, when debentures and debenture stock are issued, the loans that are received by the company become part of that company's capital. This chapter has considered some of the key principles in relation to the maintenance of the company's share capital. Importantly, however, when capital is raised by means of debentures that does not form part of the company's share capital, and so none of these principles will apply. A company is able to pay off its debts in the form of buying back any debenture stock as well as paying off loans in the ordinary way.

The registration of company charges

So far in this book the only context in which we have considered a company creating a charge is in relation to debentures, although a company could create a charge over its own property in a number of other situations, just like a human being can. The real issue in relation to company charges relates to the formalities that are necessary for such a charge to be properly created.

The charge that a company creates in relation to a debenture is an important part of the case law relating to debentures. Significantly, when a company creates a charge over its own property, such a charge will frequently require registration to be effective. The topic of registration of company charges is frequently considered to be controversial by commercial people because the maintenance of a charge requires administration and therefore involves cost. Particularly from the perspective of banks, who of course lend to companies more than any other organisation, the cost and effort involved with taking out charges, registering them and ensuring that they are maintained by the company creates a logistical headache.

The general law on charges is voluminous and so is beyond the scope of this book. There are important distinctions to be drawn between the creation of fixed charges and the creation of floating charges which also remain beyond the scope of this book. The key principles of the law relating to charges are considered in Hudson, *The Law of Finance*, 2009b, paragraphs 23–54. We shall focus specifically on company charges. Section 860(1) CA 2006 provides that:

> (1) A company that creates a charge to which this section applies must deliver the prescribed particulars of the charge, together with the instrument (if any) by which the charge is created or evidenced, to the registrar for registration before the end of the period allowed for registration.

Therefore, for a company to create a valid charge, both the particulars of the charge and the instrument that creates the charge (such as a debenture under seal) must be delivered to the registrar and registered within the specified time period. If a company fails to comply with s 860(1), an offence is committed by the company and by every officer of the company who is in default (s 860(4) CA 2006). By making this an offence, it is hoped that the company's officers will take it upon themselves to register any charge; otherwise the company would benefit if the charge was invalid because that would mean that the lender would have no valid rights to seize the property and so protect its security. The charge may be registered by 'any person interested in it': and therefore it does happen that it is the lender who seeks to register the charge because, after all, it is the lender who will benefit from a charge if the company fails to pay its debts (s 860(2) CA 2006).

The key question in many circumstances is as to the types of charge that must be registered. A large amount of litigation has centred on the question whether or not particular types of transaction create charges that ought to be registered (for example, *Re Spectrum Plus* (2005)). The types of charge that require registration are defined in s 860(7) CA 2006 in the following manner:

> This section applies to the following charges . . .
>
> (a) a charge on land or any interest in land . . .,
> (b) a charge created or evidenced by an instrument which, if executed by an individual, would require registration as a bill of sale,
> (c) a charge for the purpose of securing any issue of debentures,
> (d) a charge on uncalled share capital of the company,
> (e) a charge on calls made but not paid,
> (f) a charge on the book debts of the company,
> (g) a floating charge on the company's undertaking or property . . .

It is further defined in s 861(5) CA 2006 that the term ' "charge" includes a mortgage'. (Again, the detail of these principles is considered in greater detail in Hudson 2009b.)

The procedure for creating debentures

Much of the regulation of debt instruments like debentures is now governed by the principles discussed in Chapter 11 relating to public companies who offer their securities to the public or who have them traded on a regulated market. Issues of debentures that fall outside those regulations are governed primarily by the ordinary law of contract, given that a contract is created between the lender and borrower. Debentures, it is suggested, should be thought of as containing ordinary contractual rights and obligations between lender and borrower, distinct security rights and obligations in the form of a charge (which will only be significant if the borrower breaches its contractual obligations), and yet further rights and obligations under any trust deed that is put in place (which again will only be significant in the event of breach).

Remedies for the holders of debentures

When discussing debentures in a book of this sort, one of the first questions is always whether this subject should be discussed as part of the company's capital or as a question arising on insolvency. This is because debentures are often only significant when there is a default by the company because it has gone into insolvency or because the holders of the debentures seek to have the company wound up. Indeed one of the principal remedies for the holder of a debenture is to have the company wound up in the manner discussed in Chapter 13. Otherwise, if the company is still solvent, the debenture holders have contractual rights to sue for the payment of any amounts owing to them; and the debenture holders will also have rights under any charge that has been created to seize the charged property. As mentioned above, where there is a trustee in place, that person will have powers to act on behalf of the debenture holders.

Chapter 11

Securities regulation and raising capital

Introduction

The context of raising capital for companies

If this book were focusing, as so many company law books do, on the creation of a trading company from scratch and its struggle into life, then we would have to consider how money would be raised to fund the company's activities. When a company is first formed, the original shareholders (known as 'the members') are said to 'subscribe' to the company's constitution and consequently they pay for the first shares that are issued by the company in accordance with that constitution. From the company's perspective, this is usually the first capital that is raised by the company. The complexities that may be associated with the company's capital were considered in the previous chapter; as was discussed there, it is possible to divide between different classes of shareholders with different rights. A start-up company may also borrow money from a bank or from some other person in accordance with the ordinary law of contract, and in accordance with banking law if the lender is a bank. Bank borrowing makes no difference to a company's share capital, except that a company with too much debt may well be a poor investment for shareholders because much of the profits may be needed to pay off the debt, and except that if a company borrows money, there will be fewer shares needed to pay for the company's trading activities.

This chapter is primarily concerned with the way in which public companies raise capital by issuing shares in the securities markets. Private companies are precluded from issuing shares and bonds to the public. Therefore, this chapter will not consider private companies until the very end, because the principal focus of securities law is on issues of securities to the public at large by public limited companies. First, however, we shall consider what is meant by a 'security' by examining the use of *bonds* and *shares* in raising money from investors. (For a more detailed discussion of these regulations you should refer to Alastair Hudson, *Securities Law*, 2008a generally, or Alastair Hudson, *The Law of Finance*, 2009b, Chapters 37 to 41.)

The mechanics of securities markets in raising capital for companies

There are two ways in which a company can raise money: it can borrow it or it can issue shares. Borrowing money can be done by means of an ordinary loan or it can be done by issuing bonds. Bonds are a form of *security* that can be bought and sold like shares. In a bond issue, the company issues a large number of bonds of equal value and carrying equal rights; investors buy those bonds by paying the purchase price of the bond to the issuing company. The company is then obliged to pay interest periodically to the investors and is obliged to repay the purchase price of the bond at the end of the life of the bond. As a result, the company acquires a loan provided collectively by all of the investors, it pays interest (just like an ordinary loan) and it is required to repay the loan at the end of the contractual term (just like an ordinary loan). Significantly, the investors acquire securities in the form of the bonds: the important part about a security is that it can be sold. Therefore, the investors may be speculating on the movements in the value of the bonds on the open market, instead of simply waiting to receive interest on their 'loan' to the company. Usually there is a 'trustee' appointed over the bond issue to ensure that payment is made and to activate the default provisions in the documentation if the issuing company breaches any of its obligations. (On bond transactions in greater detail see Hudson 2009b, Chapter 35.)

Alternatively, a *public company* may decide to raise capital by issuing new shares. It is a statutory requirement that new shares must first be offered to existing shareholders (CA 2006, s 561), which means that the existing shareholders must provide more capital to the company or else those shares can be offered for sale to the public generally. If shares are bought by new shareholders, then that will dilute the voting power attached to the shareholdings of the existing shareholders, but on the other hand the company will have more capital to invest in its business. (The rules governing the allotment of shares were considered in Chapter 5.) Shares in public companies are securities in that they can be transferred (that is, bought and sold) freely in units on the open market. The remainder of this chapter considers the regulations that govern the issue of securities by public companies in the UK.

The sources of UK securities regulation

The EU roots of modern UK securities regulation

UK securities regulation is based on EC directives that set out the high-level principles on which securities regulation across the EU is based, and on technical regulations generated by the Commission that provide for the detailed rules governing securities markets. Importantly, these securities regulation directives permit member states in the EU to 'gold-plate' their own regulations, which means that the UK is permitted to create more stringent regulations than the EU minimum set out in those directives, provided that there is no discrimination between entities from the UK and entities from other member states of the EU. The UK has elected to have more stringent regulations because it is thought that this underlines the status of the UK as the foremost securities market in the EU, and consequently that issuers will be attracted to issue securities in the UK because investors know that the regulatory requirements in that jurisdiction are more demanding than anywhere else in the EU.

The EU spent a large amount of time trying to create a single market for securities and other financial services without success. The EU legislative process was too unwieldy and so tended to produce legislation that was already out of date by the time it was introduced because of the rapid pace of innovation in financial markets. Therefore, the 'Lamfalussy methodology' was introduced in 2001, which provided both that directives should only create general principles and that the Commission should produce detailed regulations, and that a Committee of European Securities Regulators (CESR) should be created to co-ordinate regulation across the EU.

The underlying policy of EU securities regulation has three main facets. First, the 'single licence' principle provides that once a securities issue has acquired regulatory approval in one jurisdiction, then it has a 'passport' to be admitted to trading in all member states in the EU without requiring any further authorisation. This is expected to move the EU towards a single market for securities and financial services because all regulations are built on the same framework principles set out in the directives and because all regulators are intended to treat the markets in their jurisdiction in accordance with the same principles. Second, this was intended to move the EU towards the harmonisation of securities regulation across all member states. However, because member states are allowed to generate more stringent regulations than their neighbours, securities regulations have by definition not been harmonised across in the EU; instead, they are merely based on the same minimum requirements, as opposed to being made the exactly same. Third, the economic aim behind securities regulation is to boost the economies of the EU by creating 'deep, liquid pools of capital' for public companies to access.

The 'pools of capital' which the policymakers have in mind include the savings and investment capital of ordinary citizens and small investors, who may be encouraged to invest in securities markets if those markets are considered to have sufficient integrity because they are regulated appropriately and because their participants behave suitably. It has tended to be more common in Continental Europe to use ordinary bank debt than securities funding, whereas the USA and the UK have tended towards a greater use of securities markets and complex financial instruments to fund economic activity. In the 21st century, EU securities regulation policy has been directed towards developing financial markets on the UK–US model, and thus on increasing the amount of capital available. Ultimately, the objective at the EU level is to enhance economic growth and increase the number of jobs in the EU through expanding economic activity, fuelled in part by the deep, liquid pools of capital that are thought to be available through securities markets. (The image of 'deep, liquid pools' is a beautiful one. If you require an image to calm yourself or to transport yourself from the travails of your ordinary life, I can suggest nothing better than imagining a 'deep, liquid pool'.)

The irony has been, of course, that the global financial crisis that began in 2007 was amplified by cutting-edge 'securitisation' structures and complex derivatives products, which were at the forefront of these securities markets that the EU wanted to tap. The impact on the EU has been seismic. Greece, Ireland and Portugal were pushed to the edge of bankruptcy so that they required bail-outs from the EU and the International Monetary Fund. Other countries such as Spain have also suffered greatly. Every country in the EU, especially the UK, has been required by modish economic thinking to cut public services drastically to repair the damage done by the financial crisis.

The implementation of the relevant EC directives into UK finance law

The underpinnings of law and the regulation of public offers of securities are to be found in the following directives: the Consolidated Admission and Reporting Directive 2001, which deals with 'listed securities'; the Prospectus Directive 2003, which deals with the requirement for a prospectus on the initial offering of securities and continuing obligations thereafter; and the Transparency Obligations Directive 2004, which deals with disclosure obligations as to the ultimate control of companies which have offered their securities to the public. The regulations are referred to in the discussion to follow as 'the EC securities directives'. There have also been important directives dealing with the regulation of market abuse and financial promotion (that is, advertising of financial products) that are also considered below. These more general directives, together with the securities directives, are referred to here collectively as 'the EC financial services directives'.

In the UK, the principal Act dealing with financial services activity is the Financial Services and Markets Act 2000 (FSMA 2000). The bulk of the EC financial services directives have been implemented by FSMA 2000. That Act has been amended on a number of occasions, and is surrounded by a huge number of statutory instruments, to implement the various parts of EU- and UK-originated legislation in this field. The principles relating specifically to the offer of securities to the public are contained in Part VI of FSMA 2000. The Prospectus Directive was, however, implemented by the Prospectus Regulations 2005 (which inserted new provisions into the FSMA 2000).

Significantly, FSMA 2000 established the Financial Services Authority (the FSA) as the principal regulator for financial services activity in the UK, in place of the ragbag of different self-regulatory organisations that had previously regulated different financial market sectors in the UK. The powers granted to the FSA by FSMA 2000 should have established it as the formidable sole regulator of financial markets in the UK. However, in practice, its preference for 'light touch' regulation of financial markets was taken to have contributed to the global financial crisis of 2007–09: in particular it was considered that the FSA had been very ineffectual in the regulation both of lending banks and of the derivatives markets, which contributed to a near systemic collapse of the financial markets in the autumn of 2008.

Among the FSA's statutory powers was the power to create the *FSA Handbook*, which constitutes all of the financial regulatory rulebooks governing financial services activity in the UK. Those rulebooks comprising the *FSA Handbook* can be viewed online at www.fsa.gov.uk. (The detail of that regulatory scheme is beyond the scope of this book, but it is exhaustively discussed in Hudson 2009b, Part 2.) For the purposes of securities regulation, there are specific subdivisions of the *FSA Handbook* known as the Prospectus Rules, the Disclosure and Transparency Rules and the Listing Rules, which implement the provisions of the EC securities directives and of the Commission's technical regulations in full. Each scheme is considered in turn in the discussion that follows.

At the time of writing, it is the policy of the coalition government in the UK to disband the FSA in 2012. Because a part of the blame for the financial crisis was placed on the ineffectual nature of having three different regulators involved in regulating banks (the FSA, the Bank of England and the Treasury), it has been decided to break the FSA up into a number of smaller pieces, thus presumably making the problem worse. Even worse than that, a large part of the FSA's responsibilities will pass to the Bank of England – which also failed to prevent the crisis – and it has been announced that key executives of the FSA who contributed to the crisis will be employed in the same roles by the Bank of England. Beyond that, the proposals for reform remain obscure. However, it does promise to be an exercise in truly staggering ineptitude, in keeping with the entire financial crisis itself. Updates on these and similar issues will be published on www.alastairhudson.com.

Prospectus regulation

The prospectus regulations in outline

The heart of securities regulation in the UK is the requirement that any company seeking to offer securities to the public or wishing to have its securities admitted to trading on a regulated market must have a prospectus relating to those securities authorised by the FSA before doing so (FSMA 2000, s 85). Failure to do so is a criminal offence. This is a very broad requirement indeed, as will emerge below, because the definition of an 'offer to the public' is an offer of a security to 'any person' (FSMA 2000, s 102B), which could theoretically include an offer to only one person. This broad requirement is then hemmed in by a few exceptions and exclusions in the statute (FSMA 2000, s 86 and Sched 2). The intention is to create a general obligation for public companies offering their securities for sale on public markets to produce prospectuses which provide the public with specified types of information mostly concerning the rights attaching to the securities, the company itself and its financial standing. These detailed contents requirements are set out in the FSA Prospectus Rules.

What is a prospectus?

A prospectus is a document that makes prescribed forms of information about securities and about their issuer available to the investing public. That prospectus will therefore constitute a series of representations on which purchasers of those securities will rely when making the decision whether or not to acquire them. The prospectus is therefore the root of any contract for the acquisition of securities, and consequently the approach of securities regulation in this context has been to prescribe the minimum contents of a prospectus for certain types of issuer and for certain types of security, and to demand the continued accuracy of that prospectus during its lifetime, so that the investing public is provided with sufficient, appropriate material with which to make informed investment decisions. This chapter, then, considers the requirements for a prospectus, the circumstances in which a prospectus will not be required, the continuing obligations associated with prospectuses and the required contents of such a prospectus.

The requirement for a prospectus and the attendant criminal offences

It is a criminal offence under s 85 of FSMA 2000 either to offer transferable securities for sale to the public in the United Kingdom or to request the admission of securities to trading on a regulated market without a prospectus in relation to that issue having first been approved by the FSA and then having been published. Consequently, a prospectus is required in relation to an issue of securities when those securities are to be offered to the public in the UK, and that prospectus must be approved by the FSA, provided that the offer is not of a type which is exempted from this requirement by the terms of the statute. Section 85(1) of FSMA 2000 provides that:

> It is unlawful for transferable securities to which this subsection applies to be offered to the public in the United Kingdom unless an approved prospectus has been made available to the public before the offer is made.

Section 85(2) of FSMA 2000 sets out a second criminal offence, which may also give rise to a civil liability to compensate for loss, in the following terms:

> It is unlawful to request the admission of transferable securities to which this subsection applies to trading on a regulated market situated or operating in the United Kingdom unless an approved prospectus has been made available to the public before the request is made.

This second offence, then, relates to the admission to trading on any regulated market: a list of such markets is kept by the FSA. It is enough that a mere 'request' for the admission of securities to trading has been made; the securities do not actually have to have been traded on that market at all. The term 'transferable securities' in this context refers to all transferable securities (including all shares, bonds and securities derivatives) except for units in an 'open-ended investment scheme' under the UCITS Directive, or government securities and other similar securities identified in Sched 11 to FSMA 2000.

An offer to the public

The purpose behind the Prospectus Directive was to capture as many offers of securities as possible and so to cast the regulatory net much more widely than had been the case before. As outlined above, s 102B of the FSMA 2000 draws the concept of an 'offer of transferable securities to the public' very broadly indeed. Section 102B(1) provides:

> For the purposes of [Part VI of FSMA 2000] there is an offer of transferable securities to the public if there is a communication to any person which presents sufficient information on –
>
> (a) the transferable securities to be offered, and
> (b) the terms on which they are offered,
>
> to enable an investor to decide to buy or subscribe for the securities in question.

Such a communication is caught if it is received by 'any person', which suggests that an offer to one person is sufficient. Consequently, the concept of an 'offer to the public' is very broad indeed in that it encompasses an offer to only one person and does not require that the offer is made to a number of people. However, the communication will be an 'offer' if there is sufficient information about the securities and the terms of the offer to enable the offeree to make up her mind whether or not to buy them. So, the mere suggestion 'would you like some shares?' would not be an 'offer' in this sense, whereas the statement 'are you interested in subscribing for ordinary shares in A plc carrying ordinary voting rights at a price of 100p each' would constitute an 'offer' if it is made to 'any person' whomsoever.

The effects of breaching s 85 FSMA 2000

There are two general effects of a breach of s 85. In criminal law terms, a failure to comply with either of these requirements constitutes a criminal offence (FSMA 2000, s 85(3)). In private law terms, a failure to comply with either of these requirements is 'actionable' on behalf of any person who suffers loss as a result of that contravention. Section 85(4) FSMA 2000 provides that a claimant is entitled to the remedies applying to actions for breach of a statutory duty in tort.

Exemptions and exclusions from the prospectus regulations

Schedule 11A to FSMA 2000 provides for three categories of securities that are excluded from the offence committed under s 85(1) FSMA 2000. In general terms they are as follows: first, government and similar securities; second, securities issued by not-for-profit organisations; third, 'non-equity transferable securities' that are 'issued in a continuous or repeated manner by a credit institution' for a total consideration of less than €50 million, or any offer with a total consideration of less than €2.5 million. There are also more specific exclusions in s 86 FSMA 2000 which relate to: offers to registered 'qualified investors', who are not thought to need regulatory protection of this sort; offers made to fewer than one hundred people, which are therefore insufficiently 'public'; large issues beyond the reach of ordinary retail investors for prices or in denominations of at least €50,000; small issues for a total consideration of not more than €100,000; or where a qualified investor acts as someone's agent.

The duty of disclosure in s 87A FSMA 2000

There is a particularly important general duty of disclosure of information in a prospectus that is set out by s 87A of FSMA 2000, which provides that:

> The [FSA] may not approve a prospectus unless it is satisfied that –
>
> (a) the United Kingdom is the home State in relation to the issuer of the transferable securities to which it relates,
> (b) the prospectus contains the necessary information, and
> (c) all of the other requirements imposed by or in accordance with this Part of the prospectus directive have been complied with (so far as those require-ments apply to a prospectus for the transferable securities in question).

In essence, the effect of this duty of disclosure is that prospectuses must contain all of the information that investors would require to make an informed assess-ment of the securities. More particularly, s 87A(1)(b) contains a requirement that the 'necessary information' be contained in the prospectus. The term 'necessary information' is defined in s 87A(2) of FSMA 2000 in the following terms:

> The necessary information is the information necessary to enable investors to make an informed assessment of –
>
> (a) the assets and liabilities, financial position, profits and losses, and pros-pects of the issuer of the transferable securities and of any guarantor; and
> (b) the rights attaching to the transferable securities.

There are therefore three significant elements to this general duty of disclosure. First, the governing principle is that the test for the type of information which should be contained in the prospectus is an 'informed assessment' test. This is not a requirement of 'reasonableness', nor is it a test based on the information that a professional advisor would require. In this regard, s 87A(4) of the FSMA 2000 provides that:

> The necessary information must be must be prepared having regard to the particular nature of the transferable securities and their issuer.

Consequently, the nature of the information required is relative to the particular securities in question and to the particular issuer. Furthermore, the prospectus must present information about the securities themselves and not generic information.

Second, the information which is required is limited to the information which would enable an investor to make an informed assessment of four things: the assets and liabilities of the issuer, the financial position of the issuer, the profits and losses of the issuer, and the prospects of the issuer. The prospectus is likely to present other types of information as well so as to market the securities to the investor base, but the types of information which are identified in s 87A FSMA 2000 are limited to these four. However, for example, a presentation of the issuer's prospects will require a broader range of information than just accounting information. Furthermore, the nature of the information that is required may also be dependent on the likely pool of investors and the complexity of the business in which the issuer is engaged, where less information will be required by expert investors or in relation to well-understood markets. Third, the prospectus must provide information as to the rights that will attach to the securities that are to be issued. In relation to debt securities or convertible securities, this may be more complex than in relation to the voting and other rights attaching to ordinary shares.

The need for comprehensible and easily analysable presentation of the necessary information

It is important that if the underlying policy of providing information to investors is to be achieved, that information must be provided in language and in a manner that is easy for all potential investors to understand. In this context, s 87A(3) FSMA 2000 provides:

> The necessary information must be presented in a form which is comprehensible and easy to analyse.

The definition of 'comprehensible' in the *Oxford English Dictionary* is that information 'may be understood'. If something may be understood by someone, then it might be comprehensible even if it cannot be understood by everyone. What is important, it is suggested, in relation to prospectuses is that an ordinary investor would be able to understand the point that is being made by the prospectus. Such an ordinary investor, it is suggested, need not be expected to be stupid but by the same token it is suggested that the prospectus should not assume anything more than a basic knowledge of investment activity, risk and financial markets in the investor base. Furthermore, very importantly, no ambiguity can be permitted in a prospectus that might conceal the true meaning of a statement in a prospectus that is otherwise encouraging investors to acquire securities; and there must also be sufficient disclosure of the basic information.

The need for a supplementary prospectus: s 87G FSMA 2000

There may be significant changes to the prospects or condition of the issuer of securities that would necessitate the publication of a supplementary prospectus to deal with those events. Section 87G of the FSMA 2000 provides that a supplementary prospectus is required to be published in the following circumstances:

> if, during the relevant period, there arises or is noted a significant new factor, material mistake or inaccuracy relating to the information included in a prospectus [or in a supplementary prospectus] approved by the competent authority.

The 'relevant period' is the time between the regulatory approval of the prospectus by the FSA and the start of trading in the securities. The requirement of 'significance' is defined by s 87G(4) FSMA 2000 to be something which is significant in relation to the categories of 'necessary information' in s 87A(2) FSMA 2000.

The required contents of a prospectus

Section s 87A FSMA 2000 provides, as discussed above, that all 'necessary information' must be provided in a prospectus so as to enable any investor to be able to make 'an informed assessment' as to the desirability of investing in those securities. However, the FSA Prospectus Rules provide for much more detailed requirements as to the contents of a prospectus. This is done in the Prospectus Rules by identifying first the categories of information which must be provided in all prospectuses (a so-called 'base prospectus'), before then providing what further categories of information must be provided in relation to specific types of securities issued by specific types of issuer. I would suggest that this is a little like a pizza: there is a generic base to the pizza in the form of the base prospectus; and then depending on the type of pizza that is required there are different toppings placed on the pizza, just as different types of security or different types of issuer require different types of information to be provided in a prospectus. These regulations are enormously lengthy and so cannot be dealt with here: reference should be made instead to the FSA website or to Hudson (2008a). What can be observed here is that the regulations do not simply provide for a 'tick box' list of the information to be provided, but rather require the issuer and its advisers to consider exactly what information is required and how that information is to be expressed and set out. The issuer is always governed by the overarching requirement that all necessary information is provided. The next section considers liability for loss caused by any shortcomings in a prospectus.

Liability for loss caused by a defect in a prospectus

Introduction

Offers of securities are a contract and the prospectus therefore constitutes a series of representations relating to the securities that are being sold. Consequently a large number of liabilities may arise under contract law and under tort law in relation to the sale of those securities if those representations turn out to have been false. The seller may be liable in contract law under the Misrepresentation Act 1967, or under tort law on the basis of the tort of negligent misrepresentation based on the principle in *Hedley Byrne v Heller* (1964), or of the tort of fraudulent misrepresentation. In this discussion we shall focus, as the cases have done, on liability in tort.

Persons responsible for the contents of a prospectus

There is an issue, first, as to the people who are responsible for the contents of the prospectus under the securities regulations. The definition of the persons responsible for the contents of a prospectus or of a supplementary prospectus are set out in Chapter 5 of the FSA Prospectus Rules. In relation to shares (encompassed within the term 'equity securities' in the regulations), the persons responsible for the prospectus are, as defined by the Prospectus Rules, 5.5.3R: the issuer; directors and those authorising themselves to be named as responsible for the prospectus; any other person who accepts responsibility for the prospectus; in relation to an offer, each person who is a director of a body corporate making an offer of securities; in relation to applications for admission to trading, each person who is a director of a body corporate making an offer of securities; and other persons who have authorised the contents of the prospectus.

In relation to securities which are not equity securities (principally bonds and securitised derivatives), the persons responsible for the prospectus are, as defined by Prospectus Rules, 5.5.4R: the issuer; anyone who accepts and is stated in the prospectus as accepting responsibility for the prospectus; any other person who is the offeror of the securities; any person who requests an admission to trading of transferable securities; any guarantor for the issue in relation to information about that guarantee; and any other person who has authorised the contents of the prospectus.

That someone has given advice in a professional capacity about the contents of a prospectus does not make that person responsible for the contents of the prospectus in itself (Prospectus Rules, 5.5.9R), unless they consent to being so named or they authorise those contents of the prospectus that are the subject of the action, and even then they are liable only to the extent that they have agreed to be so liable (Prospectus Rules, 5.5.8R).

The 'golden legacy' of strict and scrupulous accuracy in relation to prospectuses

Long before the introduction of any kind of formal securities regulation in the late 20th century, the case law had always taken the approach those preparing a prospectus relating to the issue of securities must do so with complete honesty. The 19th-century approach to the preparation of prospectuses was set out by Kindersley VC in *New Brunswick, etc., Co. v Muggeridge* (1860), 383, in the following way:

> Those who issue a prospectus, holding out to the public the great advantages which will accrue to persons who will take shares in a proposed undertaking, and inviting them to take shares on the faith of the representations therein contained, are bound to state everything with strict and scrupulous accuracy, and not only to abstain from stating as fact that which is not so, but to omit no one fact within their knowledge, the existence of which might in any degree affect the nature, or extent, or quality, of the privileges and advantages which the prospectus holds out as inducements to take shares.

This approach – that a prospectus must be prepared with 'strict and scrupulous accuracy' – in the case law was approved in many subsequent cases. So, in *Central Railway of Venezuela v Kisch* (1867), 123, Lord Chelmsford held that no misstatement or concealment of any material facts or circumstances ought to be permitted. In his Lordship's opinion, the public should be given the opportunity to judge everything that had a material bearing on the true character of the company's proposed undertakings for themselves. Consequently, it was held that the utmost candour ought to characterise the promoters' public statements.

The FSA Prospectus Rules take a broadly similar approach to the obligations of those preparing the prospectus to include all necessary information. It is mandatory under those rules to produce a prospectus when making an offer of securities to the public or when seeking admission of securities to listing on a regulated market, in that it is an offence not to make such a prospectus available beforehand. Consequently, securities regulations now impose positive continuing obligations on companies as to the full disclosure of the information required by the FSA Prospectus Rules, and are therefore not limited to 'strict and scrupulous accuracy' as to the things that the issuer chooses to mention in the preparation of the initial prospectus.

Negligent misstatement at common law

In spite of the strict approach taken in the older cases to the preparation of prospectuses, the position was limited greatly by the decision of the House of Lords in *Caparo v Dickman* (1990) to the effect that you are not liable for negligent misrepresentations if your statement was intended for a narrow group of people which did not include the claimant. It is questionable whether or not this precise approach can be supported in the particular context of offers of securities for which a prospectus is required by the FSA Prospectus Rules precisely because an issuer is required to make that prospectus (containing the information set out in the regulations) available to the public at large. Nevertheless, we shall work through the principles established by the mainstream case law chronologically before coming to the important decision in *Possfund Custodian Trustee Ltd v Diamond* (1996).

The principle underpinning liability in tort for negligent misrepresentation is contained in the decision in *Hedley Byrne v Heller* (1964). That principle imposes liability under a duty of care in the following circumstances. First, one party must seek information or advice from another party in circumstances in which the party seeking advice relies on that other party to exercise due care. It must be reasonable for the party seeking advice to rely on the other party to exercise such care and it must also be the case that the party giving the advice knows or ought to know that reliance is being placed on her skill, judgment or ability to make careful inquiry. Furthermore, the party giving the advice must not expressly disclaim responsibility for her representation.

In *Caparo Industries plc v Dickman* (1990) their Lordships stressed that the defendant's liability was limited to those whom he knew would receive the statement and would rely upon it for the purposes of a particular transaction. In that case, in relation to a takeover bid for Fidelity plc, the claimant had relied on accounts prepared by the defendants which had negligently misstated the financial position of Fidelity plc in a way which the claimant contended had caused it loss by encouraging it to take over Fidelity plc. It was held that the claimant was not within the class of people who were intended to see the accounts and therefore their loss was not foreseeable by the defendants. Lord Bridge held that '[i]f a duty of care were owed so widely, it is difficult to see any reason why it should not equally extend to all who rely on the accounts in relating to other dealings with a company as lenders or merchants extending credit to the company'. Therefore, it was held that on policy grounds a general duty of care ought not to be imposed on auditors in such situations.

Lord Bridge suggested that general statements as to the condition of company would generally be unlikely to attract liability in tort on the basis that:

> The situation is entirely different where a statement is put into more or less general circulation and may foreseeably be relied on by strangers to the maker of the statement for any one of a variety of different purposes which the maker of the statement has no specific reason to anticipate.

Therefore, his Lordship's view was that statements put into general circulation would be less likely to attract liability in tort because the maker of that statement could not ordinarily know the purposes to which his statement would be put to use by the claimant. By extension, where it is intended that only the initial subscribers for shares would be able to rely on a statement, that would mean that the makers of that statement owed no liability in tort to subsequent purchasers of shares in the after-market because the prospectus had been issued in respect of the original purchases by original subscribers for those shares.

In *Al-Nakib Investments (Jersey) Ltd v Longcroft* (1990), the claimants, who had acquired shares in the after-market, alleged that the prospectus and two interim reports issued by M Ltd contained misrepresentations as to the identity of the person who would be manufacturing the company's products. It was held by Mervyn Davies J that the prospectus had been issued for a particular purpose, namely to encourage subscription for shares by a limited class of subscribers, and that any duty of care in relation to its issue was directed to that specific purpose only. Therefore, the claimant was not entitled to sue in tort because it was not within that class of persons at whom the reports and prospectus were addressed.

A different approach to negligent misrepresentations in prospectuses

There are two cases that have taken a different approach on the facts in front of them from the cases considered above. First, in *Morgan Crucible Co plc v Hill Samuel Bank Ltd* (1991), the directors and financial advisors of the target company had made express representations in the course of a contested takeover forecasting a 38 per cent increase in pre-tax profits at a time when an identified bidder for the company had emerged. The purpose behind these and other representations had been to induce the bidder to make a higher bid because the bidder had relied on those statements. The Court of Appeal held that there was therefore a relationship of sufficient proximity on those facts between the bidder and those who had been responsible for the statements to found a duty of care in negligence.

Second, in *Possfund Custodian Trustee Ltd v Diamond* (1996) there had been a placement and subsequent purchases of shares in the after-market by the claimant. The prospectus prepared in relation to the initial placement greatly understated the issuer's financial liabilities. As a result of those financial liabilities, the issuer subsequently went into receivership. The claimant, who had bought shares after the placement, contended that it had relied on the statements made in the prospectus, that it had been reasonable for them to rely on the prospectus in that way in making their purchases, and that those responsible for the prospectus had breached a duty of care owed to the claimant as a purchaser of the shares. Lightman J held that the purpose of a prospectus at common law and under statute in English law had always been:

to provide the necessary information to enable an investor to make an informed decision whether to accept the offer thereby made to take share on the proposed allotment, but not a decision whether to make after-market purchases.

Further, his Lordship held that:

What is significant is that the courts have since 1873 (before any legislation) recognised a duty of care in case of prospectuses when there is a sufficient direct connection between those responsible for the prospectuses and the party acting in reliance (see *Peek v Gurney*), and the plaintiffs' claim may be recognised as merely an application of this established principle in a new fact situation. . . . I can find nothing in the authorities or textbooks which precludes the finding of such a duty and at least some potential support in them.

Therefore, Lightman J accepted that there could be a duty of care in these circumstances, especially given that the investors had been justified in relying on the statements made in the prospectus when making their investment decisions. It was held that such a duty of care would exist if the subsequent purchaser could show the following things: that she relied reasonably on the prospectus; that she reasonably believed that the defendant intended her to act on it; and that there was a sufficiently direct connection between the parties to make such a duty fair, just and reasonable.

It is suggested that the effect of the FSA Prospectus Rules, in relation to offers of securities to the public, is to put the entire investing public necessarily within the cognisance of any person involved in the preparation of a prospectus. Consequently, any loss caused by a negligent misstatement in such a prospectus must now be actionable in tort. Consequently, the precise *ratio decidendi* of the decision of the House of Lords in *Caparo v Dickman* has now been superseded in relation to issues of securities falling within s 85 FSMA 2000 because the makers of representations in prospectuses may not necessarily be able to hide behind a claim that their statements were intended to be made only to a limited class of buyers (so as to avoid liability to the marketplace at large).

The tort of fraudulent misrepresentation

While this discussion has proceeded thus far on the basis that the defendant may have been negligent in the preparation of a prospectus, it is perfectly possible that the company was established and a prospectus prepared solely to defraud the investing public or even just a private class of investors in relation to a private company, in which case the tort of deceit (or fraudulent misrepresentation) may apply. The measure of damages for fraud is prima facie the difference between the actual value of the shares at the time of allotment and the sum paid for them (*McConnel v Wright* (1903)). The plaintiff is entitled to recover all the actual damage directly flowing from the fraudulent misrepresentation (*Doyle v Olby (Ironmongers) Ltd* (1969)), provided that the fraudulent statement induced him to enter into that contract (*M'Morland's Trustees v Fraser* (1896)). In *Smith New Court Securities Ltd v Scrimgeour Vickers (Asset Management) Ltd* (1996), Lord Browne-Wilkinson held that where fraudulent conduct has caused loss to the claimant, the defendant is bound to make reparation for all the damage directly flowing from the transaction; and even though the entirety of that loss need not have been foreseeable, it must have been caused directly by the transaction. Furthermore, the claimant is entitled to recover as part of his damages the full price paid for the securities by him, but subject to the proviso that he must deduct from that any benefits which flowed from the transaction (including the price of the securities on sale).

In order to obtain rescission of a contract of allotment of shares on the ground that it was induced by misrepresentation, the allottee must prove, first, a material false statement of fact that, second, induced him to subscribe. If a statement is included in a prospectus, that is generally taken to be an assertion of that statement as fact. For example, in *Re Pacaya Rubber and Produce Co. Ltd* (1914), the prospectus contained extracts from a report of a Peruvian expert as to the condition of a rubber estate that the company sought to acquire. It was held that these extracts from the report contained in the prospectus formed the basis of the contract. On the basis that the company had not distanced itself from the report nor suggested that it did not vouch for the accuracy of the report, the company was taken to have contracted on the basis of the contents of the report and so the contracts of allotment could be rescinded. Therefore, the company may well seek to distance itself from supporting the accuracy of any statement made by a third party which is included in the prospectus; although, it is suggested, the very fact that a company includes a statement from a third party in a prospectus must be for the purpose of inducing investors to acquire securities and so should attach liability to the company in any event if it has induced the investor to enter into the contract.

Liability based on s 90 FSMA 2000

The Prospectus Directive required that each member state must provide for a ground for compensation for relevant defects in a prospectus, and s 90 FSMA 2000 was therefore enacted over and above the liabilities arising under common law already considered above. Section 90 FSMA 2000 therefore provides that a claimant will be entitled to compensation if she has suffered loss as a result of any false or misleading statement in the prospectus or any omission from the prospectus. A claim can be brought by anyone who has bought or who has contracted to buy any securities to which the prospectus relates. Such claimants must prove that they have suffered a loss as a result of some defect, as considered above, in the prospectus. This right to compensation, it is suggested, tallies closely with the duty of disclosure in s 87A FSMA 2000 to provide all necessary information in the prospectus. Section 90(1) FSMA 2000 provides that:

Any person responsible for [a prospectus] is liable to pay compensation to a person who has

(a) acquired securities to which [the prospectus] apply; and
(b) suffered loss in respect of them as a result of

>   (i) any untrue or misleading statement in the [prospectus];
>   (ii) or the omission from the [prospectus] of any matter required to be included by [the duties of disclosure in] section [*87A or 87B*].

There may therefore be three bases for liability here. First, in relation to 'untrue' statements, where all that is required is that the statement must have been untrue, not that there was fraud. Second, 'misleading' statements made in a prospectus where the loss is made 'a result of' the misleading statement; again there is no requirement of fraud. Third, that no material has been omitted which is required by the duty of disclosure in s 87A FSMA 2000, as was discussed above. In essence, then, it is the duty of the persons responsible for the prospectus to ensure that all statements made in the prospectus or any omissions made from the prospectus are not untrue and not misleading.

There are six possible defences to a claim under s 90 (FSMA 2000, Sched 10). The first defence applies when the defendant believed in the truth of the statement that was made in the prospectus. The second defence applies when the statement was made by an expert, when the statement is included in a prospectus or a supplementary prospectus with that expert's consent, and when it is stated in that document that the statement was included as such. The third defence requires the publication, or the taking of reasonable steps to secure publication, of a correction. The fourth defence requires the taking of all reasonable steps to secure the publication of a correction of a statement made by an expert. The fifth defence applies when the statement was made by an official person or contained in a public, official document, provided that the statement was accurately and fairly reproduced. The sixth defence applies if the court is satisfied that the investor acquired the securities with knowledge that the statement was incorrect, and therefore that the investor was not misled by it.

The general obligation to obey the Prospectus Rules

There is a general obligation imposed on all persons to whom any individual rule is specified as being applicable to obey that rule (Prospectus Rules, 1.1.3R). Further to s 91(1A) of FSMA 2000, any contravention of the listing rules opens up an issuer or any persons offering securities for sale to the public or seeking their admission to a regulated market to a penalty from the FSA if there has been either a contravention of Part VI of FSMA 2000 generally, or a contravention of the prospectus rules, or a contravention of the transparency rules. Penalties may also be imposed on directors of the perpetrator if the FSA considers it to be appropriate (FSMA 2000, s 91(2)). In place of financial penalties, the FSA may instead issue a statement of censure of the appropriate persons (FSMA 2000, s 91(3)).

Transparency obligations

Transparency obligations in outline

The keynote of investment regulation in the EU is to ensure that all investors have access to information about the nature of the securities in which they may wish to invest and about the entities that are issuing those securities. Consequently, the purpose of transparency regulation is specifically to ensure that the investing public has information about the control of voting rights in a company, as well as access to regular financial information about the company. This is how the affairs of the company are said to become transparent. Transparency regulation flows from the Transparency Obligations Directive, which was implemented in the UK by means both of Part 43 of the Companies Act 2006 (CA 2006) and of the Disclosure and Transparency Rules (DTR, which form part of the *FSA Handbook*).

The FSA Disclosure and Transparency Rules relate to dealings on 'regulated markets'. As a result of the DTR, an issuer of securities must provide 'voteholder information' and prescribed forms of financial information to the FSA. The financial information that must be provided includes annual and half-yearly accounts, and interim management statements as described in the Directive. The voteholder information relates to control of shareholders and the rights of people with rights to acquire voteholding shares in the company as well as the ownership of shares. There is also a requirement on investors to provide 'voteholder information' to the issuer for this purpose. Also required is:

- information relating to the rights attaching to any and all securities;
- information relating to any new loans or related security interests connected to them;
- information as to voting rights held by the issuer in the issuer itself;
- information as to any proposed amendments to the issuer's constitution.

Moreover, the FSA may demand information from time to time from the following types of person:

- the issuer;
- a voteholder, or a person who controls or is controlled by a voteholder;
- an auditor of an issuer or a voteholder;
- or a director of an issuer or a voteholder.

Misleading statements in relation to transparency obligations

Section 90A FSMA 2000 provides for compensation to be paid in relation to untrue or misleading statements contained in, or in relation to omissions from the reports which are required under transparency obligations. Section 90A of FSMA

2000 provides that the issuer of securities will be liable to 'pay compensation', first, on the making of an untrue or misleading statement and, second, if someone discharging managerial responsibilities within the issuer has the requisite knowledge of the nature of that statement.

The Listing Rules

The source of the Listing Rules

The securities of the principal public companies in the UK are listed on the Official List maintained by the FSA in its capacity as the UK Listing Authority. The principles governing admission to the Official List are contained in the EC Consolidated Admission and Reporting Directive, which was implemented by Part VI of FSMA 2000 and in turn by the FSA Listing Rules (created by the FSA further to powers granted to it by FSMA 2000). In essence, the listing rules are governed by the 'listing principles' which set out high-level principles governing the admission to listing, the maintenance of listing and the FSA's powers of censure in the event of a breach of those rules.

The Listing Principles

The obligations imposed on issuers by the Listing Principles require that:

- the company's directors understand their obligations;
- the issuer maintains adequate procedures, systems and controls;
- the issuer conducts its activities in relation to the Listing Rules with integrity;
- information is communicated in a way that avoids the creation or continuation of a false market in listed equity securities;
- the issuer ensures the equal treatment of all shareholders; and
- the issuer conducts its dealings with the FSA in an open and co-operative manner.

The Listing Rules are to be interpreted in the accordance with those general principles.

The Model Code on market abuse

A key part of securities regulation is the prevention of market abuse: principally the prevention of the abuse of inside information relating to securities. The Model Code is concerned with corporate governance within listed companies as it relates to the misuse of inside information. As such it is that part of the Listing Rules which seeks to prevent market abuse. Briefly put, during prohibited periods in relation to a company's securities, any insider (such as a person discharging management responsibilities or an employee in an applicable role) who wishes to deal in that company's securities must seek clearance under the procedure identified in the Rules. The aim is not only to prevent actual abuse but also to prevent the suspicion of abuse, so as to preserve confidence in the market for those securities.

Corporate governance

The Listing Rules provide that the annual financial report of a list company must include a statement as to the extent of that company's compliance with the Combined Code on Corporate Governance (Listing Rules, 9.8.6(5)R). Also to be included with this statement is a further statement as to whether or not the listed company has complied with the Combined Code throughout the accounting period or whether there are any provisions in relation to which there has not been compliance.

Applications for admission to listing

An application for listing must be made to the FSA in accordance with s 75 FSMA 2000 and, in turn, in accordance with the Listing Rules. The principles dealing with the prerequisites for admission to listing are set out in Chapters 2 and 6 of the Listing Rules. The FSA may make admission to listing subject to any special condition that it considers appropriate. To ensure that the application proceeds appropriately, all applicants are required to have a sponsor. Sponsors must be selected from a list of approved sponsors who are, in effect, expert in financial services and particularly in admission to listing, who will be required to ensure that the application and its supporting materials are appropriate and suitably prepared. The general conditions in the Listing Rules divide between conditions relating to the nature and condition of the applicant itself, conditions relating to the nature of the securities that are to be issued, including their transferability, and conditions as to the documentation that is to be provided. There are requirements that the applicant issuer must be duly authorised, have appropriate accounts prepared, have published the prescribed financial information, have appropriate management, have appropriate working capital, and so forth. There are also requirements as to the nature of the securities themselves, which relate severally to the admission of the securities to trading on a recognised investment exchange, the need for the securities to be validly issued and freely transferable, the market capitalisation of the securities, that a sufficient number of securities are to be issued, and the preparation of an appropriate prospectus.

In essence, there are different, detailed rules for the admission of different types of security to listing: therefore, the assiduous reader eager for detail will have to refer to the regulations themselves because their detail is beyond the scope of this work. The issuer is under a duty in general terms to 'take reasonable care to ensure that any information it notifies to a [regulatory information service] or makes available through the FSA is not misleading, false or deceptive' (Listing Rules,

1.3.3R). Furthermore, the issuer must not only ensure that all information is not misleading and so forth, but also that it does not 'omit anything likely to affect the import of the information' (Listing Rules, 1.3.3R). The FSA may also demand further information that it may 'reasonably require to decide whether to grant an application for admission' (Listing Rules, 1.3.1(1)R) as well as any information that the FSA considers 'appropriate to protect investors or ensure the smooth operation of the market' (Listing Rules, 1.3.1(2)R).

General continuing obligations in the Listing Rules

Listed securities must continue to be admitted to trading on a regulated market and at least one quarter of their number must remain in public hands. Significantly, issuers of securities must publish accounting information, including a preliminary statement of their annual results as soon as such a statement has been approved. Publication of information must be done through a recognised information service once those matters have been agreed with the company's auditors, together with any decision as to the payment of a dividend (Listing Rules, 9.7.2R).

The preservation of the integrity of securities markets requires that inside information be published as soon as appropriate. The aim is to prevent people from abusing that information by using that information to make profits on dealings in securities, as considered below in relation to insider dealing. In general terms, the FSA Disclosure and Transparency Rules oblige an issuer to notify a recognised information service 'as soon as possible' of any inside information which 'directly concerns the issuer' (DTR, 2.2.1R), unless the issuer (on its own initiative) considers the prevention of disclosure to be necessary to protect its own 'legitimate interests' (DTR, 2.5.1R). The FSA created a code to specify what sorts of behaviour may constitute market abuse in the FSA Market Abuse Rulebook.

Furthermore, the fourth Listing Principle in the Listing Rules provides that: '[a] listed company must communicate information to holders and potential holders of its listed equity securities in such a way as to avoid the creation or continuation of a false market in such listed equity securities'. There are a number of matters about which a listed company must make disclosure either at the time of making an issue of securities or on a continuing basis thereafter. The keynote here is the avoidance of the creation of a 'false market' in the equity securities at issue.

FSA powers of punishment: discontinuance and suspension of listing

The FSA has four separate powers under FSMA 2000 to prohibit or suspend or otherwise control securities transactions, including:

- the power to discontinue or suspend listing;
- the power to suspend or prohibit an offer of transferable securities to the public;

- the power to suspend or prohibit admission to trading on a regulated market; and
- the power to suspend trading in a financial instrument on grounds of breach of the FSA disclosure rules.

The FSA may publish a statement of censure if an issuer of transferable securities, or a person offering transferable securities to the public, or a person requesting the admission of transferable securities to trading on a regulated market, has failed to comply with its obligations under any provision of the securities regulations stemming from Part VI of FSMA 2000.

Insider dealing and market manipulation

If securities markets require ordinary investors, there must be a perception among those investors that financial markets have integrity and that insiders are not able to gain an unfair advantage over other investors, or to manipulate the prices on markets. Therefore, there are offences relating to 'insider dealing' and 'market manipulation' that arise under Part V of the Criminal Justice Act 1993 (CJA 1993) and FSMA 2000 respectively. Insider dealing has been criminalised in part so that there is no inequality of bargaining power between parties, where one party has inside information which the other party could not have, as well as to preserve a perception of market integrity among investors.

Further to s 52 of the CJA 1993, the principal offence of insider dealing arises where the defendant 'deals in securities that are price-affected securities in relation to the information'. The principal offence in s 52(1) is expressed in the following terms:

(1) An individual who has information as an insider is guilty of insider dealing if, in the circumstances mentioned in subsection (3), he deals in securities that are price-affected securities in relation to the information.

The elements of the offence are therefore as follows. First, the offence is committed by an individual (that is, a human being). Second, that individual must have information as an 'insider', as is defined below, as opposed to having that information in any other way. Third, the individual must 'deal' in securities. Fourth, the securities in which the individual deals must be 'price-affected securities', as is defined below. Fifth, those securities must be price-affected securities 'in relation to the information', and not coincidentally price-affected due to some other factor. Thus, the insider must be dealing in relation to information which itself affects the price of the securities. Sixth and furthermore, these activities must be performed in the circumstances set out in s 52(3):

(3) The circumstances referred to above [in s 52(1)] are that the acquisition or disposal in question occurs on a regulated market, or that the person dealing

relies on a professional intermediary or is himself acting as a professional intermediary.

The offence is therefore focused on buying or selling securities on the basis of inside information. There is a dealing in securities whenever securities are bought or sold by the defendant, or where she procures such a purchase or sale. Information is 'inside information' if it relates to securities, if it is specific and precise, if it has not been made public, and if it had been made public it would have had a 'significant effect on the price' of securities. The defendant must be an insider who has access to this information by virtue of her employment or other professional duties, or must have the information from an 'inside source'. It is also an offence to encourage another person to deal in securities or to disclose information improperly to another person.

The offences of market manipulation are contained in ss 397 and 398 of the FSMA 2000. Market manipulation arises where a 'statement, promise or forecast' is made which is 'misleading, false, or deceptive in a material particular'. The purpose of the offence is to prevent the market value of shares being artificially raised or lowered by the dissemination of rumour and so forth. It is an offence to encourage someone to enter into transactions or to refrain from transacting as a result of such a statement.

Chapter 12

Mergers and takeovers

Introduction

This short chapter considers three types of corporate restructuring: takeovers, mergers and internal reorganisations. The key theme of this book has been that companies are tools that are used to achieve particular ends. So when their human users change their minds about those goals or the best means of achieving them, they can change their companies too. There may be changes in regulation or taxation; or the businesses within a group of companies may wax or wane, suggesting a need to keep some businesses and sell off others; or a multinational group of companies may need to reorganise itself as there are changes across those jurisdictions in which it is based. The forms of restructuring that are considered here are mergers, takeovers and internal reconstructions, although others are possible.

Mergers

The types of merger

The ideal form of merger is a combination of two pre-existing companies into a new company. Often in practice, mergers do not improve the business of either company because there is frequently bad blood as one person from each pre-existing company typically vies for the one remaining post of that type in the new company, and as the managements of the two pre-existing companies vie for control of the new company. Therefore, mergers sometimes happen by 'absorption', as one company is absorbed into another. In complex mergers of two *groups* of companies, there will be many different subsidiary companies that must seek to meld together. Often in a complex merger of that sort it is possible to maintain the distinct personality of some of the pre-existing business units by locating them within subsidiaries of the resulting merged holding company in a similar form to the position before the merger. In such a situation, for most of the employees of those business units there will be little change to their working lives, apart from a new logo on the wall and the occasional change in the corporate bureaucracy.

The fundamental principles underpinning mergers in the Companies Act 2006

When a merger of public companies is proposed, a draft of the proposed terms of the scheme must be drawn up and adopted by the directors of the emerging companies, including information about the share exchange ratio between the companies (s 905 CA 2006). The draft terms must be published at least one month before the company meetings that will discuss and vote on them (s 906 CA 2006). The directors must prepare a report for the vote at that meeting which explains the effect of the merger on the company and which sets out the legal and economic grounds for the proposal (s 908 CA 2006). There must also be a report prepared on the proposal on behalf of both companies, including commentary on questions of valuation (s 909 CA 2006). Also an 'expert's report must be drawn up on behalf of each of the merging companies' which considers the share exchange ratio (s 909(1) CA 2006). The draft proposed terms must be delivered to the registrar and notification of receipt must then be published in the Gazette at least one month before any meeting of that company to discuss the proposed merger (s 906 CA 2006). The merger must then be approved by a 75 per cent majority of the shareholders in all of the merging companies (s 907 CA 2006). Similar provisions apply, *mutatis mutandis*, in relation to divisions of companies (s 920 CA 2006) instead of mergers.

Mergers in insolvency

Under s 110 of the Insolvency Act 1986, a company which is in voluntary winding up procedures may transfer or sell the whole or part of its business or property to another company. In the case of a creditors' winding up, the liquidator's authority for the passage of the special resolution must come either from the court or from the company's liquidation committee. Typically in such circumstances, a meeting of the transferor company is summoned in order to pass resolutions for reconstruction or amalgamation. At the meeting, resolutions are passed for the voluntary winding up of the company, the appointment of a liquidator, and the grant of authority to the liquidator to enter into an agreement with the transferee company on the terms of a draft submitted to the meeting. The agreement typically provides that the transferee company shall purchase the assets of the transferor company.

A sale or arrangement under s 110 is binding on all members of the transferor company, whether they agree to it or not (s 110(5)). However, s 111 provides that a member who did not vote in favour of the special resolution, and who also expressed dissent from it in writing to the liquidator, may require the liquidator either to abstain from carrying the resolution into effect or may require the purchase of his interest at a price which is to be determined by agreement or arbitration. The liquidator must satisfy all creditors' claims in relation to the transferor company. However, the creditors may petition for a compulsory winding up order if they consider that their position will be prejudiced.

Takeovers

The sources of takeover regulation

The Takeover Code ('the Code') is the principal source of regulations governing takeovers in the UK. It is drawn from the Takeover Directive (2004/25/EC): those principles are now implemented into UK company law by Part 28 of the Companies Act 2006. The Code relates to takeover offers for companies whose shares have been admitted to trading on a 'regulated market', such as the London Stock Exchange. The Financial Services Authority maintains a list of 'regulated markets'. In effect the Code relates to the UK's largest public companies.

The underlying aims of the Code are to ensure an orderly mechanism for the emergence of a bid for a takeover and the conduct of a takeover. A takeover, broadly, is where one company acquires the shares in another company. Even the rumour of a takeover has a great effect on the price of a company's shares. Therefore, the preservation of an orderly market requires that takeover rumours be controlled. Yet it is a difficult thing to amass sufficient shares to take over a public company, which means that the risk of rumours developing as these shares are acquired is high. The Code aims to provide the market, including shareholders in a company that is the subject of a takeover bid (a 'target company'), with information about the strategy of a person (a 'bidder') who is considering making a bid to take over a target company. The purpose of the Code is to require that announcements be made when direct or indirect holdings of shares cross particular thresholds, and then that the conduct of the takeover bidding process itself follows the procedure set out in the Code.

The Takeover Panel ('the Panel') administers the Code, as it administered previous regulations on takeovers. It is generally considered that the Takeover Panel performs a good job of overseeing takeovers despite being a form of self-regulatory body. Its functions are governed by statute for the first time since the enactment of the CA 2006. It has the power to regulate takeovers granted to it by s 942 of the CA 2006, albeit that it is obliged to co-ordinate with the Financial Services Authority.

The statutory powers of the Takeover Panel

The Panel is empowered to give rulings on 'the interpretation, application or effect of rules' (s 945 CA 2006), to give directions to restrain actions in breach of the rules, and to compel compliance with the rules (s 946 CA 2006). In the exercise of its powers, the Panel has the power to compel the production of documents and other information (s 947 CA 2006). In keeping with the need to maintain the secrecy of the process, the Panel must keep anything learned through this process confidential (s 948 CA 2006), and the disclosure of any such information constitutes an offence (s 949 CA 2006).

The Panel operates through an executive committee. Watkins LJ expressed the function of the Executive in *R v Panel on Takeovers and Mergers, ex Parte Guinness plc* (1988) in the following terms:

It is the executive which takes the lead in examining the circumstances of takeover bids and, if thought necessary, referring them to the Panel for consideration and adjudication according to the rules. Almost daily it is called upon to give advice and rulings, which mostly are accepted. . . . It acts as a sort of fire brigade to extinguish quickly the flames of unacceptable and unfair practice.

If the Executive or either of the parties considers that the matter is serious enough, it will be referred to the Panel for a full hearing. The Takeover Code provides that the role of the Executive is as follows:

The day-to-day work of takeover supervision and regulation is carried out by the Executive. In carrying out these functions, the Executive operates independently of the Panel. This includes, either on its own initiative or at the instigation of third parties, the conduct of investigations, the monitoring of relevant dealings in connection with the Code and the giving of rulings on the interpretation, application or effect of the Code.

The Executive therefore has an important role to play in relation to the interpretation of the Code and in relation to communication with market participants. It is also the Executive that provides the first line of implementing the Code in individual cases.

The Takeover Code

The purpose of the Takeover Code

The underlying purpose of the Code is as follows:

The Code is designed principally to ensure that shareholders are treated fairly and are not denied an opportunity to decide on the merits of a takeover and that shareholders of the same class are afforded equivalent treatment by an offeror. The Code also provides an orderly framework within which takeovers are conducted. In addition, it is designed to promote, in conjunction with other regulatory regimes, the integrity of the financial markets. The Code is not concerned with the financial or commercial advantages or disadvantages of a takeover. These are matters for the company and its shareholders. Nor is the Code concerned with those issues, such as competition policy, which are the responsibility of government and other bodies.

The Code has developed as a consensus position among those in the 'market' as to the best means of treating takeovers.

The general principles underpinning the Code

The Code operates on the basis of the six following 'general principles':

1. All holders of the securities of an offeree company of the same class must be afforded equivalent treatment; moreover, if a person acquires control of a company, the other holders of securities must be protected.

2. The holders of the securities of an offeree company must have sufficient time and information to enable them to reach a properly informed decision on the bid; where it advises the holders of securities, the board of the offeree company must give its views on the effects of implementation of the bid on employment, conditions of employment and the locations of the company's places of business.

3. The board of an offeree company must act in the interests of the company as a whole and must not deny the holders of securities the opportunity to decide on the merits of the bid.

4. False markets must not be created in the securities of the offeree company, of the offeror company or of any other company concerned by the bid in such a way that the rise or fall of the prices of the securities becomes artificial and the normal functioning of the markets is distorted.

5. An offeror must announce a bid only after ensuring that he/she can fulfil in full any cash consideration, if such is offered, and after taking all reasonable measures to secure the implementation of any other type of consideration.

6. An offeree company must not be hindered in the conduct of its affairs for longer than is reasonable by a bid for its securities.

These principles mirror those in Article 3 of the Takeover Directive.

The way in which the initial offer is made and then publicised

The keynote of the Code is the provision of accurate information to the market-place as soon as practicable at each stage in the process to prevent the destabilising effects of takeover rumours on a company's share price and the concomitant drain on management time. (All references to 'rules' are to provisions in the Code.) The initial offer is to be made to the board of the offeree company. There must be 'absolute secrecy' (rule 2) in relation to the offer, which is to be treated as 'inside information' for the purposes of the regulation of market abuse by the Financial Services Authority. Transmission of this information to other people can only be made if it is 'necessary'. The primary responsibility for making an announcement rests with the offeree once it has been notified of the offer. Under Rule 9 there is a mandatory obligation to make an announcement of the takeover offer.

To prevent the lead-up to the offer being allowed to drag out (thus increasing the likelihood of a leak and taking up management time), the offeree may impose a timetable on the offeror requiring it to firm up the detail of its offer. To prevent time-wasting, the Code requires that the offeror should only announce a firm intention to make an offer once it has given the matter the most careful and responsible consideration. If so, a 'firm announcement' may be made which trails the formal offer to come, or equally importantly the offeror may make a statement of an intention not to make an offer. In the meantime, the board of the offeree company is required to obtain competent independent advice on any offer (rule 3), which in turn involves expense.

The regulation of market abuse and the criminalisation of insider dealing

There is a great deal of concern in the Code about the dangers of insider dealing on the basis of sensitive information before it becomes known initially, and then also about any further developments subsequently (whether as to likelihood of an offer or the revocation of the offer, or the price at which the offer will be effect, and so on). Rule 4.1 provides that 'no dealings of any kind in securities of the offeree company by any person . . . who is privy to confidential price-sensitive information concerning an offer or contemplated offer may take place' between the time when an approach is first contemplated and when an announcement is eventually made. Rule 4.2 then prohibits dealings in securities by the offeror and 'concert parties' in restricted periods. Rule 4.6 prohibits the unwinding of a borrowing or lending transaction in relation to relevant securities by the offeror, the offeree company, their associates, their connected advisors, their pension funds and so forth, without the consent of the Panel.

Once the possibility of an offer has been published, there remains the ongoing problem of the offeror slowly building up a holding of shares in the offeree and rumour leaking out into the market about its rate of progress. Therefore, rule 5 of the Code identifies the types of public disclosure that the offeror must make as its accumulated shareholding in the target company crosses a series of thresholds. Instead of being limited to direct ownership of the shares in a company, the Code refers to bidders having 'control' of a given a proportion of the voting rights derived from the company's shares either directly or indirectly, which prevents secret holdings of shares from being built up. And instead of ownership, the acquisition of interests in shares is sufficient (such as acquiring an option to buy shares). The Code defines 'control' of a company as constituting a 30 per cent holding of voting rights in that company.

Importantly, once the offeror acquires 30 per cent of the voting rights (that is, shares carrying voting rights) in the target company, then it is required to make a 'mandatory offer' by rule 9 of the Code. This prevents the takeover process from drifting on for too long. It is interesting to note that the shareholding in a public company is ordinarily so widespread that acquiring a holding of 30 per cent of its shares is taken to be sufficient to constitute control of the target company for this purpose. That mandatory offer must be made to the holders of any class of equity share capital, and is usually in the form of a price which the offeror is prepared to pay for each share to acquire the remaining shares in the company sufficient to take practical control of the target.

The reference in the Market Abuse Directive to 'inside information' is a reference to:

> information of a precise nature which has not been made public, relating, directly or indirectly, to one or more issuers of financial instruments or to one or more financial instruments and which, if it were made public, would be likely to have a significant effect on the prices of those financial instruments or on the price of related derivative financial instruments [such as options to buy or sell shares].

The FSA market abuse code is known as the 'Code on Market Conduct', incorporating the categories of market abuse set out in s 118 of the Financial Services and Markets Act 2000, and provides that 'dealing on the basis of inside information which is not trading information' constitutes market abuse.

Moreover, insider dealing involving such information is a criminal offence under the Criminal Justice Act 1993. In essence, it will constitute market abuse to deal on shares in a company on the basis of inside information and will invite the intervention of the regulator. The FSA now has the power to investigate and prosecute the criminal offences under the 1993 Act, whereby it is a criminal offence if the defendant deals in securities that are price-affected securities in relation to inside information. So s 52(1) of the 1993 Act provides that:

> An individual who has information as an insider is guilty of insider dealing if, in the circumstances mentioned in subsection (3), he deals in securities that are price-affected securities in relation to the information.

The elements of the offence are therefore as follows. First, the offence is committed by an individual, not a company. Second, that individual must have information as an 'insider' – such as being a director or employee of the company, or being someone who acquires the information through their job or profession, or being someone who has acquired the information from such a source – as opposed to having that information in any other way. The information must have been precise information which would significantly affect the price of the securities and which had not been made public. Third, the individual must 'deal' in securities, for example by buying or selling them. Fourth, the securities in which the individual deals must be 'price-affected securities'; that is, securities the price of which would be affected significantly by the information in question. Fifth, those securities must be price-affected securities 'in relation to the information' and therefore their price must not be coincidentally affected by some other factor. Sixth, these activities must be performed in the circumstances set out in s 52(3):

> The circumstances referred to above are that the acquisition or disposal in question occurs on a regulated market, or that the person dealing relies on a professional intermediary or is himself acting as a professional intermediary.

So, the securities must be the securities of a public company quoted on a regulated securities market (as discussed in Chapter 11).

It is also an offence to do either of the two following things, such that further to s 52(2) of CJA 1993 the following arises:

> An individual who has information as an insider is also guilty of insider dealing if –
>
> (a) he encourages another person to deal in securities that are (whether or not that other knows it) price-affected securities in relation to the information, knowing or having reasonable cause to believe that the dealing would take place in the circumstances mentioned in subsection (3); or
>
> (b) he discloses the information, otherwise than in the proper performance of the functions of his employment, office or profession, to another person.

Therefore, it is an offence to encourage another person to deal in the securities (for example, getting your spouse to deal on your behalf) or to tip someone else off as to possibilities offered for the price of those securities by the confidential information which you know (for example, telling your friends at the golf club so that they can make a profit). The detail of these offences is beyond the scope of this book, but it is analysed in Chapters 12 and 14 respectively of Hudson (2009b).

The listing rules

The listing rules, as discussed in Chapter 11, will apply to takeovers of listed companies over and above the Code.

Directors' duties in relation to takeovers

As discussed in Chapter 7, the general duties of directors have been developed in relation to takeovers in particular because hostile takeovers so often divide management and the shareholders, and in consequence the directors need to demonstrate that they are acting in the best interests of the company, that they are exercising their powers for a proper purpose and that they have no conflicts of interest. Those principles are now enshrined in s 170 *et seq.* of CA 2006, as discussed in Chapter 7.

Corporate reorganisations

So far this chapter has considered mergers and takeovers, which are at the exciting end of the spectrum and which earn investment banks huge sums in fees during boom times. However, there are more mundane aspects to corporate reorganisations that reflect a change in the corporate strategy.

In a complex corporate group – for example a bank – there will be a huge number of subsidiary companies conducting different parts of the business, and even subsidiary companies which are created solely to acquire computers or to acquire vending machines or to handle the catering. More importantly, the business operated through different business units and in different jurisdictions will typically be conducted through different subsidiaries, so that any business that has to be subject to regulation in Ruritania is kept distinct from business relating to Ruritania but which the bank wishes to keep beyond regulation by the Ruritanian authorities. The *Salomon* principle is particularly useful in these contexts for regulatory avoidance. So, if the regulatory regime changes in Ruritania so that the bank decides to move parts of its business out of that country, then it is likely that it will want to reorganise which businesses are assigned to which of its subsidiaries.

The same would be true of a manufacturing company that wanted to sell off the business unit that manufactured widgets if the company had decided to concentrate on manufacturing twidgets. The company would transfer the widgets business into a separate subsidiary company so that it could be sold off more easily. There is no merger or takeover here. Instead there may be a division of assets between companies, or a reorganisation of the shareholding of pre-existing subsidiary companies.

In Chapter 6 we considered how corporate groups operate in this fashion. A part of restructuring requires that the appropriate votes and resolutions of each company involved are passed at the appropriate meetings, with the appropriate reports and so forth having been circulated. Therefore, internal restructurings require careful administrative handling. As we have discussed in this book, there is a reasonable part of company law which is entirely bureaucratic, and based ultimately on the need to provide information in part by keeping records of decisions.

Chapter 13

Corporate insolvency

Introduction

This is the thirteenth chapter of this book. Traditionally for the British the number 13 is considered to be unlucky. So, it is appropriate that this is the chapter in which we consider corporate insolvency and the way in which companies, in effect, die. During bad times in the economy, many company law practitioners find themselves specialising in insolvency as companies are wound up and others go into insolvency. This chapter divides between a discussion of winding up solvent companies and, its principal focus, winding up insolvent companies. Because winding up for solvent and for insolvent companies come to similar endpoints, it makes most sense to consider both forms of winding up together, even though this chapter is primarily concerned with insolvency. There are a number of methods of insolvency for companies, each of which depends upon the circumstances of the business and seeks to achieve different goals. We shall consider each in turn. Before we turn to the detail of the legislation, however, we need to consider some of the broader issues that relate to insolvency.

The sources of insolvency law

The principal legislation governing law on corporate insolvency is the Insolvency Act 1986 (IA 1986) and the Enterprise Act 2002. The IA 1986 reforms the position that was previously based on the Bankruptcy Act 1914. The principal legislation is reinforced by the Insolvency Rules 1986 and other regulations.

The role of insolvency law

In some jurisdictions around the world there is no concept of insolvency. It is a very difficult thing to maintain a capitalist marketplace without having an idea of corporate insolvency. Insolvency should be thought of in this context as being the way in which companies die. If companies are indeed to be considered as a legal person with distinct personality, then it must be possible for them to die. The logic of capitalism dictates that when a trading company ceases to be effective and cannot pay its debts when they become due, that company should be terminated and moved out of the way so that more productive organisations can spring up in that space. The metaphor that is commonly used is that of 'clearing away the dead wood'. If you think of a forest, if dead trees are allowed to remain standing for too long, they will prevent new vegetation from growing on the forest floor. Therefore, it is better to clear the dead wood out of the way. If companies were allowed to continue in perpetuity when they had ceased to be useful or where they had ceased to be profitable, it would be in nobody's interests. Jurisdictions in which companies were incapable of dying in this way simply lingered on in a form that is referred to by some economists as 'zombie capitalism'.

In this chapter, once we have considered the concept of insolvency itself, we shall consider five different legal methods of dealing with insolvency: receivership, liquidation, administration, voluntary arrangements with creditors, and winding up. Each of these different concepts presents a different way of dealing with an insolvency. In essence, the thinking is this. If a company can no longer continue in existence, it is as well simply to wind that company up and terminate its existence. However, it is common that companies will still have some assets which could be sold off, and therefore a liquidator would be required to take charge of the process of gathering as much money as possible for the assets which are left, so that that money can be distributed among the creditors. Alternatively, it may be that the business could continue to earn some money if it carried on for a short period with a skeleton staff under the control of an administrator. This would still mean that the company would ultimately be terminated but that some money could be realised in the meantime instead of simply shutting up shop and walking away. Alternatively still, the company may be involved in an activity which will still earn a cash flow, for example receiving rent from leases over land, and therefore a receiver may be put in place to collect that future cash flow, even though other activities involving the company would cease. Yet further, it may be that the company and its creditors can come to some agreement as to the way in which a struggling company will operate in the future for the benefit of all involved, on the basis of a formal agreement which is solemnised in court.

Very importantly, the principal focus of insolvency law is on the position of the company's creditors. When a company goes into insolvency, everyone's concern is to maximise the amount of money and other property which can be made available to satisfy the obligations which are owed to the company's creditors, and consequently to attempt to minimise the loss which might otherwise be suffered by those creditors. One of the cardinal principles of insolvency law is the *pari passu* principle, which means that all creditors are to be treated in a way which gives no unfair advantage to one creditor (or to one class of creditor) over any others. Literally, each is to proceed 'in equal step'. However, secured creditors (that is, creditors who have some proprietary rights already in existence in relation to the company's property, or some other such rights recognised by insolvency law) will have recourse to their property first, whereas unsecured creditors (that is, creditors who have no such protection or security) must look to the general assets of the company once the claims of secured creditors and others identified in the legislation have been satisfied. In my own mind I always picture the unsecured creditors as being a miserable bunch standing in the rain outside locked factory gates, hoping beyond hope that there will be some assets left in existence which will pay off some of the debts which are owed to them. Among the unsecured creditors of a trading company will often be its trade creditors who have supplied it with goods, customers who have pre-paid for goods, its own employees who are waiting for their wages, banks which have lent it money without security (perhaps by means of an overdraft), and so on.

Difficulties with knowing when a company is insolvent

It is a surprisingly difficult thing to know when a complex trading company is insolvent. Classically, a company will be insolvent when it is unable to pay its debts as they become due. That means, on the date when a company is due to make a payment to a creditor it is unable to do so because it has insufficient assets to meet that claim. The larger the company, the more difficult it may be to know whether or not that is the case. In very large organisations there will be a huge number of suppliers, customers and others who will owe money to the company and be owed money by the company, so often deciding whether or not a company is insolvent will depend upon whether or not the banks are prepared to maintain its lines of credit.

What this means for company law is that it is not always a legal question whether or not the company has gone into insolvency. Instead, it is likely to be an accountancy question as to whether or not the company should be considered to have become insolvent on account of its cash flow position, its unprofitability, a reduction in its net worth, and so forth. Because companies are abstract entities, they obviously do not die in the same way that physical beings die. Therefore, it is a matter of professional art to know whether or not a company should be considered to be insolvent, and therefore dead. Given the room for manoeuvre which this suggests, the various types of insolvency law procedure which we shall consider in this chapter are concerned with the way in which a business may be kept ticking over either in the hope that it can be brought back to life (for example, if a buyer for the business can be found, or if the general economy improves) or so that its creditors can be brought into the best possible position in the circumstances.

Insolvency practitioners

The orderly administration of insolvency law depends in large part upon the activities of qualified insolvency practitioners. To qualify as an insolvency practitioner one must follow the route into qualification set out in s 390 CA 2006. Most insolvency practitioners are members of a recognised professional body, such as an accountant or a solicitor.

The different legal mechanisms for dealing with insolvency

The layout of this discussion

The remainder of this chapter will consider each of the legal mechanisms for dealing with and conducting the insolvency of a company in an order which begins with the most benign form of action ('the voluntary arrangement') and then works through the other options, until we come finally to the ultimate winding up of the company, which causes it to cease to exist in law. There are two principal policies in this legislation which emerge from the discussion to follow: first, an effort in recent legislative innovations to attempt to keep the company solvent or to restore it to good health if at all possible and, second, an underlying policy to achieve equal treatment for different types of creditor as part of an effort to realise as much money or other property as possible to meet as much of the claims of all creditors as possible in the circumstances. The economic impact of insolvency law can be enormous if failing companies are extinguished too lightly, because the effect on other parties who have traded with that failed company will be that they suffer loss, thus causing a ripple effect out through the economy. Nevertheless, the logic of a market economy requires that moribund or failed businesses must be extinguished so that fresh economic activity and investment can flourish elsewhere.

Voluntary arrangements

One of the innovations in the IA 1986 was the creation of procedures for voluntary arrangements with a company's creditors. This had the effect of making it possible to bind creditors into an agreement that would avoid a liquidation. Under s 1 of the IA 1986, any one of the directors or a liquidator or an administrator of a company may make a proposal to the company for what is known as a 'composition' to create a 'voluntary arrangement'. In essence, a voluntary arrangement sets out a scheme whereby the business will be operated so as to meet the creditors' claims over time. It seeks to prevent the creditors from seeking a formal liquidation or administration of the company provided that the company itself performs its obligations under that voluntary arrangement. In that sense, it is a sort of negotiation between the company and its creditors that reorganises the manner in which the company will meet its obligations over time. The thinking is that it is in the best interests of the company, all of its stakeholders and its creditors if the company's business is given the opportunity to get past a difficult period and return to profitability. It is certainly in the interests of the economy more broadly if businesses can get past any stumbling blocks and succeed in the longer term. Often it will be one creditor, or a small group of creditors, who may jump the gun and seek to enforce their rights ahead of everyone else, attempting to have the company wound up when that is in no one else's interests. Significantly, voluntary arrangements will not bind creditors who refuse to be bound by them from the outset, as discussed below.

The directors must prepare a proposal that explains why they consider the voluntary arrangement to be desirable and why it would be in the interests of the company's creditors (Insolvency Rules 1986, rule 1.3(1)). This proposal must also set out details of the company's assets, whether they are charged in favour of secured creditors, and all matters that are appropriate to assist the creditors to reach an informed decision about the proposal (Insolvency Rules 1986, rule 1.3). It is now possible for the directors of the company to obtain a moratorium of 28 days to prevent creditors from seeking an order for the administration of the company while the voluntary arrangement is in effect (IA 1986, S. 1A (1)). The legislation imposes controls on the amount of credit that a company can obtain during the moratorium period, the amount of the company's property that can be dealt with during that period, and so forth. Meetings of the creditors are then convened to consider the voluntary arrangement. Significantly, the voluntary arrangement cannot affect the rights of a secured creditor unless the secured creditor consents to that. Ordinarily, any creditor who had notice of those meetings and was entitled to vote at those meetings will be bound by the voluntary arrangement (IA 1986, s 5). Creditors may, however, challenge the voluntary arrangement within 28 days on the basis that it will unfairly prejudice their interests or that there has been some material irregularity in relation to the meetings (IA 1986, s 6). If there is no challenge, the voluntary arrangement will be put into effect, and overseen by a supervisor (IA 1986, s 7 (2)). The supervisor may apply to the court for directions, but otherwise must oversee the carrying into effect of the voluntary arrangement and must maintain accounts and records.

Receivership

The law of receivership has existed for many centuries, and reaches outside corporate insolvency. Importantly, receivership is dependent upon a court order due to its significant ramifications for the operation of a company. There is a general right to petition the court for the appointment of a receiver under section 109 of the Law of Property Act 1925.

In essence, the role of a receiver in this context is to receive any rents or profits which flow from a business (or particular items of property) and to apply them in meeting the needs of creditors. To do this, a receiver takes possession of the company's property with a view to realising its profits. The principal objective of a receiver is to protect the position of creditors, usually creditors with some security interest in relation to the company. For example, a receiver may be appointed in relation to a specific item of the company's property that is being used to provide a charge for one specific creditor (such as a debenture holder). The receiver would not be running the entirety of the business in this example but rather would be protecting the interests of that creditor in relation to that specific property. (In relation to a floating charge such a person is referred to as an administrative receiver further to s 72A IA 1986.)

Clearly, it will be very inconvenient for a company to have a receiver appointed in relation to some of its property because that will interfere with the operation of its business and other activities. Therefore, the question arises: in which circumstances will a receiver be appointed by the court in relation to a company? There are four clear situations in which the case law has recognised a right to appoint a receiver. First, where amounts are owed by the company under a debt to its creditor and those amounts are in arrears (*Re Crompton* (1914)). Second, where the company is in the process of being wound up (*Wallace v Universal Automatic Machines* (1894)). Third, where it is considered that there is some risk that the security provided by the company (for example, a charge over property) may be impossible for the creditor to realise (for example, if that property was at risk of being destroyed or taken out of the jurisdiction) (*London Pressed Hinge Company Limited* (1905)). So, if the judge is convinced that chattels or monies which were subject to a charge might be taken out of the jurisdiction, that would justify the appointment of a receiver (*International Credit and Investment Company v Adham* (1998)). Fourth, where it appears unlikely that legal proceedings would acquire discharge of the debt for the creditor (*Soinco v Novokuznetsk Aluminium* (1998)).

The appointment of a receiver has a number of legal consequences. For example, if property is held subject to a floating charge in favour of the creditor, that floating charge crystallises and becomes a fixed charge (*Re Crompton* (1914)). Significantly, for the purposes of company law, the powers of the director to manage the company can be suspended in favour of the receiver in relation to the property that is subject to the receivership (*Gomba Holdings UK Ltd v Homan* (1986)). Whether the company continues in business will depend upon the circumstances. In the event that the receiver takes power over all of a company's property, there is authority to the effect that all of the employees of the company may be dismissed (*Griffiths v Secretary of State for Social Security* (1974)); although, it is suggested, that that would depend upon whether the receipt of the profits involved required the continued presence of those employees. In some circumstances, the receiver may be deemed to act as the agent of the company in carrying on the company's business to protect the creditor's interests, although the receiver is not otherwise considered to be the agent of the company. Whether or not a receiver is to be deemed to be the agent of the company, as opposed to acting solely on behalf of creditors, is a matter of interpretation from case to case (*OBG Ltd v Allan* (2007)). Generally, when the business of the company continues under receivership, its communications must make clear that it is acting in this way. The remuneration received by the receiver itself is fixed by agreement (IA 1986, s 36).

Insolvency proceedings can be complex, not least because different creditors will have different views as to the best way of operating a company through troubled times. Therefore, it may be that a winding up of the company is sought even when it is under receivership. If the winding up of a company does commence, the receiver will no longer be an agent of the company (if that was the case in any event) and will lose the power to bind the company, except in relation to any property in relation to which the receivership took effect.

Administration

The role of an administrator is to attempt to save a company from being wound up or to attempt to acquire as much money as possible in the disposition of the company's assets (often involving selling them). When the concept of administration was first introduced in 1985, its purpose was to provide an alternative to the company being liquidated. Much of insolvency law policy after 1985 has been concerned with finding alternatives to the insolvent winding up of companies by means of other legal procedures which would both protect the interests of creditors while also keeping the company's activities hirpling on so far as is possible or desirable. Given the enormous amount of flexibility in these sorts of cases, court orders always used to be required to ensure that no individual creditor or group of creditors is being disadvantaged, although today other procedures have been developed in parallel, as discussed below. The law in this area was most recently updated by the Enterprise Act 2002.

The policy underpinning the most recent reforms to the law and administration in the Enterprise Act 2002 was the need to provide a company 'in financial difficulties with a breathing space in which to put together a rescue plan or, alternatively, in providing a better return to creditors than would be likely in a liquidation' (*Productivity and Enterprise: Insolvency – A Second Chance* (Cm 5234 The Insolvency Service, 2001)). Consequently, the administrator of the company has three principal objectives: to rescue the company as a going concern, or to achieve better results for the company's creditors than liquidation would provide, or to realise property (that is, by selling it) so that the proceeds can be distributed among secured or preferred creditors (Enterprise Act 2002, s 3(1)). Therefore, where possible the company is to be kept alive as a viable business rather than killed off. Importantly, an administration appointment will terminate after one year. Therefore, an administration is not considered to be anything like a permanent arrangement.

Administration can be put in place in one of three ways. First, if an application is made to the court for an administration order, an administrator may be appointed by the court on the basis that the company is either unable to pay its debts or that it is likely to become unable to pay its debts. The decision of the High Court in *Colt Telecom Group plc* (2002) took the view that in deciding whether or not a company is unable to pay its debts, it must be proved that it is more probable than not that the company will be in that position. The courts must be convinced that the administration presents a real prospect of success. Second, the holder of a floating charge may appoint an administrator of the company, provided that the charge was created before 15 September 2003. Third, the company or its directors may appoint an administrator, which would be something that the directors may consider if the company was in dire financial straits and in need of outside help. In all events, the administrator must be a qualified insolvency practitioner (Enterprise Act 2002, s 390). All administrators are officers of the court (Enterprise Act 2002, s 390).

The appointment of an administrator has a number of legal consequences. Once an administration order has been made, any winding up petition against the company is dismissed, except where required in the public interest (s 124A IA 1986), or where the Financial Services Authority has presented a petition to wind up the company (under s 367 of the Financial Services and Markets Act 2000). Furthermore, creditors are prohibited from seeking to enforce their security over the company's property during the administration. In this way, the underlying intention of providing a breathing space for the administrator to put the company back to rights is obtained.

The administrator has a variety of powers that are granted to it by Schedule 1 of the Insolvency Act 1986. An administrator has the power to take possession of the company's property and to take any legal proceedings in relation to it which seem appropriate; a power to sell and otherwise deal with the company's property; a power to borrow money, and to grant security over the company's property; a power to appoint professional advisers; a power to act in legal proceedings on behalf of the company; a power to use the company's seal; and so forth. The general power that is perhaps most significant for an administrator is the power to carry on the company's business and to organise its capital. Nevertheless, in practice the administrator is required to produce proposals for the operation of the business so as to achieve the best possible outcome for the creditors of the company.

The administrator also bears a number of duties. As was mentioned above, an administrator is an officer of the court and therefore is responsible to it. An administrator has an overarching obligation to act 'honourably' (*ex parte James* (1874)). Beyond that, the administrator bears typical fiduciary duties to take custody and control of all relevant property and to take reasonable care to obtain a proper price of any property sold. An administrator does not generally bear direct obligations towards creditors, however, without expressly assuming such an obligation (for example, *Kyrris v Oldham* (2003)).

One remarkable development in the practice relating to insolvency administration has been the development of so-called 'pre-pack' administrations. Where the intention is to sell a company and its business, to improve the attractiveness of that package it has become common for the administration procedures to have been finalised before the formal insolvency stage is reached but, significantly, also before the sale is effected. In this way the purchaser has much less involvement with the complexity of the administration procedure. At the time of writing, it appears that these procedures have not been considered four-square by a court in this jurisdiction, although the practice has received indirect approval by way of *obiter dicta* in some cases. A discussion of the structures is set out by Sandra Frisby in *Charlesworth's Company Law*, paragraph 28-018.

Winding up a company

For our purposes, we think of a company as being wound up in two different circumstances: when the company is solvent and when the company is insolvent. The members of a solvent company may decide that that company is of no further use to them and therefore they may decide to bring its existence to an end. Alternatively, if a company is insolvent, an order for its winding up may be made further to a petition to the court. In the insolvency law jargon we must distinguish between a 'voluntary' winding up (either by the members or by creditors) and a winding up by the court, and a winding up on the 'just and equitable' ground.

Winding up on the just and equitable ground

We have already considered the possibility of winding up a company on the 'just and equitable ground' under s 118 IA 1986 in Chapter 8. In that discussion we considered the termination of the company in *Ebrahimi v Westbourne Galleries* (1973), in which a business had been set up by two men and operated as a quasi-partnership between them, until one of the men brought his son into the business and there was a falling-out – which meant that the father and son tried to block the other participant's access to any of the company's assets, with the result that the House of Lords considered it appropriate to wind the company up so that its assets could be distributed among the three men and a difficult situation brought to a close. In the next section we shall be considering the other bases on which a winding up may be effected, which are generally more procedural in nature than the broad basis on which a winding up may be ordered under s 118.

The other grounds on which the court may order a winding up

Section 122 IA 1986 provides that a company may be wound up on a number of different bases. The most significant bases for our purposes are:

- if the company has resolved to be wound up by a special resolution which is approved by the court;
- if the company is unable to pay its debts;
- or if the company does not commence business within a year after it is incorporated, or if it suspends business for an entire year.

To take the last situation first, if the company is not conducting a business, then there is no particular objection to the company being wound up. If the company has been unable to pay its debts, then insolvent winding up is appropriate – and that is what we shall consider first.

Liquidation and insolvent winding up

The role of a liquidator is to wind up a company and to terminate it. At that point, the company ceases to exist, as though it had died. In that sense, the company is 'liquidated'. Under s 123 IA 1986, a company will be deemed to be unable to pay its debts if one of a number of events has occurred:

- a creditor who is owed more than £750, has served a demand on the company in the prescribed form and the company neglects to make its payment on that debt; or
- execution on a judgment is returned either wholly or partly unsatisfied; or
- the court is convinced the company is unable to pay its debts as they become due on the basis of evidence laid before it; or
- it is proved to the court's satisfaction that the company's liabilities exceeded its assets (including any future and contingent assets and liabilities).

These last two examples of a company being unable to pay its debts as they become due will depend upon expert accountancy and other evidence which, as discussed earlier in this chapter, may not be as clear-cut as one might otherwise expect. In relation to the first two situations, the existence or the precise terms of the act or other claim may be at issue between the parties, of course. It is also the case that, particularly in poor economic circumstances, companies seek to make payments as slowly as possible to protect the amount of cash that is held in the business. Nevertheless, permitting a petition on the basis of persistent failure to pay debts is generally considered by the courts to be appropriate, so as to encourage the proper performance of such payment obligations in commercial life generally (for example, *Taylor's Industrial Flooring Ltd v M H Plant Hire Manchester Ltd* (1990)).

The people who may present a petition for a winding up are several (s 124 IA 1986), including the company or its directors, a creditor or creditors of the company, a contributory or the Secretary of State. The most common method for commencing a petition for winding up on the basis of insolvency is for a creditor to present a petition to the court. Typically, it will be an unsecured creditor who brings this action, although a secured creditor whose security has turned out to be in some way deficient may also serve a petition: otherwise a secured creditor is more likely to be able to rely on its security, instead of needing to present a petition to wind the company up. Petitions are served by the Secretary of State on the basis that it is considered to be in the public interest that a particular company should be wound up, at which point it is a question for the court to decide whether or not it is just and equitable to order the winding up (s. 124A IA 1986).

The insolvency regulations set out the procedure for the advertising and distribution of an insolvency petition before it is served on the company and heard in court. The purpose of these regulations is to give the company an opportunity to regularise its affairs before the petition is heard, and also to bring the matter to the attention of any parties who may have a stake in the company's future.

When the matter reaches court, the court may be asked to appoint a provisional liquidator (s 135 IA 1986). The provisional liquidator then takes all of the company's property under its control (s 144 IA 1986), from which point no other legal proceedings may be commenced (s 130(2) IA 1986). At the hearing of a petition, the court may do any one of a number of things: grant the petition, refuse to grant a petition but make an interim order, adjourn the proceedings or dismiss the petition outright (s 125 IA 1986). An order to wind the company up on this basis is generally referred to in the jargon as a 'compulsory winding up'.

If an order to wind up the company is made, then all dealings with the company's property are frozen, and any purported transfers of the company's property would be void (s 127 IA 1986). The company is no longer the owner of its own assets, and instead all of the company's property is held on the terms of a statutory trust (*Ayerst v C&K (Construction) Ltd* (1976)). Once a winding up order is made, a number of maudlin consequences follow. For example, any employees of the company are deemed to have been dismissed on the basis that the company ceases to exist, which has been the position since the 19th century (*Chapman's Case* (1866)).

It is then the role of the liquidator to terminate the company and to organise the dispersal of its assets in an orderly fashion. Strictly speaking, it is the official receiver (a public functionary) who acts as the liquidator until another liquidator is formally appointed. The official receiver is to enquire into the causes of the failure of an insolvent company, and into any other relevant dealings. This may bring to light evidence of fraudulent or wrongful trading (which was discussed in Chapter 3). A new liquidator may be appointed as a nominee of the creditors or else the matter may be passed through the Secretary of State. It is common for meetings of creditors and other people entitled to a contribution from the assets of the company to appoint a liquidation committee to oversee the various aspects of the liquidation, which may be required in complex insolvencies. The liquidator must then get the company's property in and identify all people who are required to contribute to the assets of the company and to organise meetings of creditors and such contributories. Ultimately, the liquidator owes its duties to the company itself as its agent.

Voluntary winding up

By contrast with the compulsory winding up just considered above, there is also a 'voluntary winding up' in which a solvent company may be terminated. Some companies are wound up on this basis because their articles of association have specified that the company will last only for a given period of time, although it is more common for the company in general meeting to have decided by special resolution that the company should be wound up. Typically, this latter decision is made because the members no longer consider that the company has a useful purpose to perform. This is all the more likely with companies which are not trading companies, although it may be that a trading company ceases to have any useful activities as well.

A voluntary members' winding up is managed by the members themselves without the need for any meeting of creditors and so forth. Consequently, it may only take place in relation to a company that remains solvent. A declaration of solvency in relation to the company is therefore required (s 89 IA 1986). The members then appoint a liquidator, who acts as the agent of the company in the winding up once that liquidator has been appointed by the company in a general meeting (s 91 IA 1986). At periods identified in the legislation, the liquidator must summon general meetings of the company and lay before it an account of the conduct of the liquidation.

A voluntary creditors' winding up requires a meeting of the company's creditors after a resolution for the winding up of the company has been passed by the company in general meeting. The legislation identifies the procedural requirements for giving notice of such a meeting. This type of voluntary winding up is managed both by the members and by the creditors, although control will ultimately be in the hands of the creditors, because in this form of winding up the creditors as well as the company have the right to nominate a liquidator. Furthermore, the creditors may appoint a liquidation committee of no more than five people to act alongside the liquidator in the conduct of the liquidation. Again, the liquidator acts as the agent of the company. At the end of the liquidation of the company, the liquidator must call final meetings of both the company and the creditors to present an account of the liquidation.

From the commencement of the winding up process (when the resolution is passed) the company must cease to carry on its business (unless some activities are required in the best interests of the company during the winding up) (s 188 IA 1986), and no dealings with the company's shares can be made without the agreement of the liquidator (s 88 IA 1986). The good news is that a voluntary winding up does not require that the employees of the company be deemed to have been dismissed immediately, although the liquidator may choose to dismiss those employees. The liquidator in this sense acquires all power in relation to the affairs of the company, subject to any complaint by the members, the creditors or any other person with *locus standi* to bring any matter before the court.

As was discussed in Chapter 8 of this book, the rights of shareholders were described as being rights to participate in the company democracy if their shares carried votes, a right to receive a dividend only if one was declared, a right to deal with their shares in accordance with the company's constitution and (significantly for present purposes) a right to participate in the company's property only in the event of a winding up of the company. The shareholders' rights to participate in the company's property, however, will be subsidiary to a number of other factors.

First, there would generally be expenses incurred in the winding up itself, including paying for the remuneration of the liquidator. The liquidator will then apply the company's property to pay off the 'preferential debts' of the company (including secured creditors); after that any property secured by way of a floating charge, then the claims of unsecured creditors, and then any property remaining is distributed among the shareholders, further to the IA 1986 and the insolvency regulations. Ultimately, the company is dissolved and ceases to exist. For this to happen, a final meeting of the company must take place and a notice of that meeting must be given by the liquidator to the registrar of companies.

Chapter 14

Corporate social responsibility

Introduction

Thinking about corporate social responsibility

What is corporate social responsibility, and what has it got to do with company law?

This chapter considers the idea of 'corporate social responsibility' that has emerged in recent years as part of a growing consciousness about the effects of globalisation on the developing world, about the degradation of the environment and about the role of companies in those things. For our purposes, the advocacy of the concept of corporate social responsibility by a range of different groups and organisations worldwide is an assertion that companies should take responsibility for the impact that they have on the world beyond the issues that company law has traditionally recognised.

The growth of concern about corporate social responsibility is really part of a political movement which is seeking to call 'big business' to account for its impact on the human and physical world around it. Advocates of corporate social responsibility would argue that this has a direct effect on company law because it is company law which ought to be providing a means by which companies can be called to account and a way in which companies can be required to introduce different factors into their decision-making beyond simple concerns with profit. It could be argued that it is the narrowness of company law, with its focus solely on the internal concerns of the triangle of director, shareholder and company, which prevents a different culture from developing within companies. Or, put another way, a change in company law to require different considerations to be taken into account (as with s 172 CA 2006) and also to give other people *locus standi* to enforce their concerns from outside the company would have a direct effect on the way in which capitalism operates through companies.

Corporate social responsibility is concerned with a range of issues which are not traditionally considered to be anything to do with company law: concerns with poverty across the world, with the degradation of the earth's natural environment, with the rights of employees, with the impact of business on the built environment even in developed countries, and with more general ethical problems bound up with modern capitalism. It is because multinational companies now have enormous political influence and financial power which reach across national boundaries that attention has shifted to the responsibility of companies for those sorts of environmental, social and ethical concerns, although the same underlying moral questions about the responsibility of the human participants in companies relate to all forms of company. Particularly in the 'developing world', large companies from the developed world are responsible for large amounts of economic activity which is conducted at low cost and often using methods which would be unacceptable or even illegal in the developed countries in which those goods or services are ultimately sold.

Moral responsibility: do no harm should be the whole of the law

As a consequence of this growing awareness of corporate social responsibility (CSR), radical company lawyers have begun to shift their attention from the technicalities of UK company law or US company law to the impact of the activities of companies beyond the reach of those particular jurisdictions. So, for example, it is possible under UK company law to avoid responsibility being imposed on the UK holding company (let alone its management or shareholders) for the harm and terminal illness caused to mineworkers in South Africa if the South African business is conducted through a South African subsidiary, because that South African subsidiary is treated as being separate from the holding company in the UK (as in *Adams v Cape Industries* (1990)). This technical approach to the company law question of the liability of the holding company for the acts and omissions of the subsidiary overlooks any moral question about the harm that is caused to others by the unthinking operation of that business.

The starting point of most philosophical discussions about personal liberty and social justice begin with John Stuart Mill's idea that we should do no harm to others (Mill 1859). It is generally taken as being the lodestar around which those debates should run: we can do what we please so long as we do no harm to others. In the *Adams* case what was permitted was the repatriation of profit to the UK to be distributed among the shareholders but no repatriation for the harm that was caused by the source of those profits. The visceral, physical risks of earning those profits were taken by the South African workers, but the profits themselves were taken by the shareholders. Here we see that company law is itself bound up with the possibility of taking a benefit without the burden of responsibility. Limited liability may be a shield, but it can also cut like a sword. It allows shareholders to be free to take profits, but it also allows others to suffer harm.

By denying any transference of liability on the *Salomon* principle, company law is locked into comparatively arid, technical debates of the sort we considered in Chapters 3 to 12 of this book because it is shielded from the need to consider any other sorts of questions. Consequently, company law overlooks the immorality of repatriating profits to a UK holding company while denying any liability for the appalling harm caused to the employees of a South African subsidiary. Allegations of a similar sort have been made against oil companies drilling for oil outside their home jurisdictions and against multinational manufacturing companies of all stripes. (Visit www.corporatewatch.org.uk for an analysis of the questionable activities of some companies.) The problem is not limited to one sector. Instead, large public companies have realised that they can make most profit if they simply own brands, patents and other intellectual property, if they transmit production overseas such that they owe no direct legal duties to the human beings who do the work, and then if they sell their products into avaricious developed markets at huge profit. (We shall consider this aspect of globalisation below.)

Much of the concern about corporate social responsibility from the perspective of a corporate lawyer is with the public relations position of the company, and also with the broader liability of companies for negligence and so on. Sometimes the causes of some of this harm can be more difficult to identify, and sometimes company law adds to that complexity. Let us take an example of hideous suffering. The leak of methyl isocyanate gas at the Union Carbide plant at Bhopal, India in December 1984 caused truly horrific injuries to thousands of people, which still impact severely on children born to those people even today. Estimates suggest nearly 4,000 deaths immediately after the leak and more than 40,000 people seriously harmed, many of them dying over time from their injuries. The latest court decision considering the attribution of legal responsibility for this incident was decided in June 2010, 26 years after the incident. As Professor Peter Muchlinski explains, and very significant for our purposes, much of the litigation concerning the attribution of fault was mired in controversy about whether or not the holding company of Union Carbide controlled the subsidiary that operated the Bhopal plant sufficiently closely to justify the attribution of responsibility to it (Muchlinski 1999, p 325 *et seq.*). Union Carbide has a website devoted to telling its story relating to the incident (www.bhopal.com). Interestingly, a part of that story relates to the fact that the business has since been sold to another company and the trading name of the business changed. Nevertheless, Union Carbide reportedly settled out of court with the Indian government for US$470 million (for example, *Time* Magazine, accessed 27 February 2001).

Of course, stories of corporate harm of this sort are not entirely unusual: the BP Deepwater Horizon oil spill off the coast of Louisiana and the Exxon Valdez oil spill in Alaska are two environmental disasters which spring to mind. BP nevertheless posted fourth quarter profits of US$4.4 billion in 2009 during a recession in which the steeply rising price of oil was exacerbating the economic harm suffered by the world economy after the financial crisis. While ordinary people were suffering real hardship as a result of corporate activity, companies like BP were generating record-breaking profits. It is in relation to these sorts of issues that corporate social responsibility is significant. The aim of this chapter is to present an overview of some of the debates as they relate to company law. We will do this by beginning with some important milestones in economic theory and business theory, before moving into the modern corporate social responsibility debate and its interaction with debates about company law.

The observations of Berle and Means about the power of companies

At the outset it would be useful to build a bridge between well-established company law theory and modern corporate social responsibility thinking. Berle and Means began their well-known account of *The Modern Corporation and Private Property* (1932) with the following observation:

> Corporations have ceased to be merely legal devices through which the private business transactions of individuals may be carried on. Though still much used for this purpose, the corporate form has acquired a larger significance. The corporation has, in fact, become both a method of property tenure and a means of organizing economic life. Grown to tremendous proportions, there may be said to have evolved a 'corporate system' – as there was once a feudal system – which has attracted to itself a combination of attributes and powers, and has attained a degree of prominence entitling it to be dealt with as a major social institution.

What Berle and Means observed here was that the company (by which they mean trading companies) is not simply a private institution which is used to organise assets, in the way that a family trust had previously been a private mechanism for holding property which did not need to interact with the outside world. By contrast, large trading companies and investment entities are the building blocks in which economic life is organised in our societies, to the extent that the company has generated a 'corporate system' in which it is companies that dominate economic life and that have consequently become incredibly powerful. In our modern company law it is recognised that companies have human rights (which is surely the doctrine of legal personality taken too far) and it is a fact of life that corporate donations to politicians in the Western world have a tremendous influence on the destination of political power.

The distance between institutional shareholders and responsibility for the business

Berle and Means are most renowned for identifying that companies are divided between ownership (by the shareholders) and management (principally by the directors and by middle management). What we have today (as identified by Parkinson 1993) in relation to public companies is a cadre of institutional investors that own the majority of the shares in the country's largest public companies, with much of the remainder being held by individuals or small-scale investors. There is necessarily a division between management and shareholders of this sort because the investors are only interested in receiving a stream of profit from the company, and are not interested generally in the direction of the company except to the extent that it is needed to generate profits.

The power to run the company is therefore located entirely with the management, except to the extent that management needs to generate sufficient profits to satisfy the major shareholders. Therefore, management is effectively *internal* to the company (running it from day to day, working in the building, strategising for its future), whereas the shareholders are *external* to the company because they are not so involved (in public companies). (One obvious exception to this would be employees who have shares in the company for which they work.) What this means is that shareholders, possibly more so than ever, have no direct connection with the moral shortcomings of the company. Institutional investors themselves are operated by corporations with corporate policies that somehow exist apart from the human beings who work there.

Nevertheless, the social place of large public companies has become such a mainstream concern that every major company from car manufacturers to banks now tries to market an image of corporate social responsibility and has formal statements of corporate social responsibility goals published. This chapter considers how corporate social responsibility has acquired a central status in the operation of many large public companies, and asks whether this commitment to corporate social responsibility is genuine or simply a cynical marketing ploy.

First, as a preface to a survey of the corporate social responsibility literature, we should begin by understanding the theoretical debates that underpin whether or not having companies concern themselves with social responsibilities is in any event a good thing.

The two poles of political economy and CSR

The best way to begin an examination of corporate social responsibility is by setting out the two competing poles of political theory during the 20th century which impact on the debate about CSR: the right-wing, free market approach which decries CSR thinking and the left-wing approach which celebrates CSR thinking as a bulwark against the worst excesses of capitalism. At the left-wing end of the spectrum is the egalitarian position (most easily identifiable with Marx, and with Klein and Bauman in this chapter) and at the other end is the free market position (most easily identifiable with Friedrich Hayek and Milton Friedman). We shall begin with the free market ideology, which underpins so much of modern company law.

The free market ideology

The argument in a nutshell

In essence, the free market position maintains that companies should seek the maximum profit in a market free from governmental interference, that a company's money and other assets should not be spent on anything other than the expenses of the business, and that companies should have no moral responsibilities beyond the conduct of business and the pursuit of profit (Friedman 1962). On this model, the pursuit of profit is considered to be a good in itself. In essence, those who argue for CSR are concerned to achieve other goals (social, ecological or otherwise) or to resist the pursuit of profit for profit's own sake.

The road from serfdom

One of the luminaries of this ideology and one of the foremost commentators at the right-wing end of the economic-political spectrum was Friedrick Hayek, who wrote *The Road to Serfdom* (1944) in the shadow of the cataclysmic Second World War, which claimed millions of lives. That was a war which wrote 'genocide' into the global lexicon of man's inhumanity to man. Hayek saw the danger of ideology being pursued to its illogical conclusions. He influenced the Chicago School of free market, libertarian economists and also the right-wing Thatcher and Reagan administrations in the UK and the USA respectively. The UK (and Chile) became a crucible for experiments with free market, monetarist economics. Hayek was frightened of socialism. His focus was specifically on the 'socialism' that gave rise to National Socialism (that is, 'Nazism' or 'fascism', which actually opposed itself to communism) in Germany in the 1930s. For Hayek, 'socialism means slavery'. In truth, the Nazis and the communists of the USSR had little in common ideologically apart from a propensity to murdering millions of people in pursuit of ideological goals.

Instead of the slavery into which Hayek believed socialism would drive humanity, he wanted freedom for individual human beings so that they could work productively in ways that would be economically successful or unsuccessful depending on the effect they had on the world around them:

> The gradual transformation of a rigidly organised hierarchic system into one where men could at least attempt to shape their own life, where man gained the opportunity of knowing and choosing between different forms of life, is closely associated with the growth of commerce.

For Hayek, European history is about 'freeing the individual from the ties which bind him'. In turn, European development is about 'the inventive faculty of man' and 'the individual innovator'. It is 'the individualist tradition which has created Western civilisation'. This chimes in with the focus of company law on the entrepreneur. This focus on freedom is central to the approach of company law that directors are constrained only by the company's constitution, but otherwise are to seek the success of the company in a way that the courts will not question unless they oppress shareholders in the manner considered in Chapter 8. Notably, effects on other people will not be similarly actionable. Company law focuses instead on the triangle of director, shareholder and company.

The free market hypothesis

The clearest central statement of free market economic principles was set out by Milton Friedman in *Capitalism and Freedom* (1962). At its root the hypothesis is that truly free markets will find the most efficient means of conducting economic activity at the lowest prices and in an atmosphere of enlightened self-interest. The economic role of a company is solely to generate profits for the shareholders, and it is this clarity of focus that, it is said, will cut out waste and inefficiency. For Friedman there are three central tenets of economic policy: big government is bad, so that the State should only exist to protect its people from force and fraud; entrepreneurial businesspeople must be set free; and regulation must be removed. From a company law perspective, in tune with this model, the sole objective of directors is to make money for their shareholders. The purpose of the State on this model is: 'to protect our freedom both from the enemies outside our gates and from our fellow-citizens: to preserve law and order, to enforce private contracts, to foster competitive markets', but nothing more.

Consequently, CSR is seen by such economists as being a waste of resources that ought to be applied to the generation of more profit by companies. In the USA, in the celebrated case of *Dodge v Ford*, the Dodge brothers were shareholders in Henry Ford's business who objected when Ford decided that his company was earning more profit than was necessary and so decided to distribute some of the 'excess' profits into the community. The Dodge brothers successfully sued him on the basis that profit should be applied solely in the furtherance of the business or be distributed among the shareholders, but could not be spent on collateral purposes like good works in the community. (The position under English company law was always more permissive than that: *Hutton v West Cork Railway* (1883).) In the modern world, of course, companies do spend money on the arts and on charitable giving, at the very least because it is considered to be good public relations which will in turn lead to greater profits.

Difficulties with the free market hypothesis

What emerges from this pen portrait is that ethics do not have a place here, aside from the avoidance of fraud. Profit is all that matters. We have already considered in Chapter 13 that pure free market economics is content to allow bad businesses to fail, but this can sometimes seem a little callous. For example, Friedman considered that the devastation wrought on New Orleans by Hurricane Katrina was an 'opportunity to reform the education system' in that city, even though the more pressing problem seemed to be that massive numbers of people were made homeless or killed. Equally worryingly, Friedman also advised the dictator Pinochet after his revolution in Chile how to put right-wing economic theory into effect in tandem with his extreme right-wing social policies and the torture of those who opposed him. As Galeano (1983) put it: 'How can this inequality be maintained if not through jolts of electric shock?'

Furthermore, if Friedman was right about the benefits of deregulation, then why did the banks fail in 2008 after all the liberalising deregulation of the 1990s and the 2000s? Another part of the free market model is the hypothesis that free markets are necessarily efficient markets. The financial crisis suggests that this cannot always be true, if it is ever actually true. Even Alan Greenspan, former Chairman of the Federal Reserve Bank of New York and architect of monetary policy in the US boom years during the 1990s, admitted, when giving evidence to a Congressional committee in October 2008, in a great reversal of faith, that markets cannot be trusted to act in this manner. Cooper has traced the inefficiencies in even deregulated financial markets (the closest model to Friedman's perfect world) and demonstrated that they do not act efficiently over time, frequently acting irrationally in the short term (Cooper 2008).

The shock in this doctrine

What is little understood within the 'free market versus egalitarian' economic debates is the impact that the insurgent theory of free market economics has had in various countries around the world. In her book *The Shock Doctrine* (2007), Naomi Klein traces the influence of Milton Friedman and the other economists from 'the Chicago School' – both the academics and their students from the University of Chicago – as they brought free market economics to deeply troubled countries like Indonesia, Chile and Argentina. In each of these countries there were vicious right-wing coups in which thousands of people were killed or 'disappeared' by the insurgent regimes. Milton Friedman expressed the view in the Preface to the 1982 edition of *Capitalism and Freedom* that:

> Only a crisis – actual or perceived – produces real change. When that crisis occurs, the actions that are taken depend on the ideas that are lying around. That, I believe, is our basic function: to develop alternatives to existing policies, to keep them alive and available until the politically impossible becomes politically inevitable.

This demonstrates a political purpose underpinning the development of ideas of free market economics. What Friedman was arguing was that (writing in 1982) it would be impossible to introduce their policies in ordinary circumstances. Instead, what was needed was a crisis so severe that their programme could be introduced. In countries like Chile, after President Allende had died in office during a military coup instigated by General Pinochet, there was just such a crisis in which the brutal civil war that erupted after the coup provided a smokescreen for the introduction of radical economic change on the Chicago School model, which was imposed by former graduates of the University of Chicago who had been educated there by virtue of a special programme started by the university to disseminate this ideology around the world.

Klein tells how a pre-prepared plan for economic overhaul had been written in advance for each of these countries, like a playbook for economic revolution, so that it could be implemented in the immediate aftermath of the coup. In Bolivia, its imposition took only a couple of weeks while political opponents were held in custody. On this programme, all state supports would be stripped away, completely free markets introduced, and inflation crushed. In each circumstance, Chicago University professors were consulted or employed by these regimes. The Suharta regime in Indonesia and the Pinochet regime in Chile were bywords for human rights abuses and large-scale killing of the populace, even after their anti-democratic coups had been carried through. This did not deter the Chicago School from interacting with them. What this disturbing litany of abuses indicates is that capitalism does not need democracy to succeed and that some free market purists are perfectly prepared to work without it. The history of this sort of capitalism is that the original laboratory experiments for it were in the aftermaths of anti-democratic military coups. This is perhaps why it does not always work in democratic societies without hiccoughs.

The egalitarian position

The egalitarian position is concerned with the need for greater equality around the world and in particular with the removal of the social injustices which are caused by multinational companies in developing countries. The trend towards globalisation in the world economy seemed profitable on its surface – because producers were able to reduce their costs and so increase their profits – but underneath there was a pattern of unconscionable labour practices, poverty wages and other abuses in many circumstances.

Globalisation and corporate social responsibility

One of the most trenchant critics of the activities of many multinational companies has been Naomi Klein, who in her book *No Logo* (2000) analysed the changes in the business practices of the largest companies. So, to take an easy example, the Nike business model is no longer that of the 'vertically integrated' organisation with all grades of employee from management to the factory floor working in one organisation, but rather it is an 'outsourced' model where management maintains the brand (with most of its costs being on advertising) and all other activities are outsourced to other people. The Levi Strauss business model sacrificed hip, Californian customer loyalty for closing plants in the USA and shipping all production offshore. The Levi Strauss subsidiary in the USA is now avowedly focused on 'brand management' but not on making clothes. The American managers are concerned with finding cheap producers and advertising the products, but little else. Klein argued that 'as the old jobs fly offshore, something else is flying away with them: the old-fashioned idea that a manufacturer is responsible for its own workforce'.

The concern about production methods for many textiles and footwear manufacturers is the use of sweatshop labour in this outsourcing of production. Among the problems that are well documented with many of the world's best-known clothing brands, and which are set out in detail by Naomi Klein, are the following:

- the use of underage workers;
- extremely long shifts with little time off in ways which would be unlawful in the UK;
- the use of compulsory dormitory living on-site for workers, sometimes involving 'three day shifts, sleeping under machines';
- unsafe and insanitary living and working conditions;
- no employment rights for workers;
- no unionisation;
- no sickness benefits;
- short-term contracts and poverty wages;
- charging employees for food and board;
- using local workers for only low-skill assembly work.

The purpose of these arrangements is solely to earn the 'brand company' huge profits by cutting production costs. The brand companies are of course concerned about the reputation cost to them of this sort of labour; therefore, they develop corporate statements of minimum standards in franchisees' factories, although commonly it is found that the production companies breach these agreements.

A liquid world of global brands

Professor Zygmunt Bauman identifies a form of *liquid modernity* in the expansion of global capitalism, in which the old social building blocks have melted away and in their place is a constantly shifting sea of choices, opportunities and risks (Bauman, 1999). Multinational companies in particular are free of the ties of geography because they are able to remove themselves from any particular location without any fetters. Not for them the ties of place, nor the burdens of creating direct contractual links with the workforce employed by the franchisee in that place. For example, manufacturers of footwear and leisurewear such as Nike have established 'lines of flight' from their producers (Bauman 1999). Nike has achieved this by owning only one really valuable item of property, namely the trade mark over the 'swoosh' logo that is sewn onto clothing and footwear. Nike does not need to own factories, nor employ a manufacturing workforce. Instead, Nike develops templates for its branded goods and then has them manufactured by producers in developing nations where labour costs are much lower than in the developed world.

Consequently, Nike is able to maintain very high profits but in a way that means it has little direct link to the human beings who make its products. Nike can therefore extract itself from relations with its franchised producers and move production elsewhere. There is no need to interact with trade unions, or take care of healthcare costs, or anything of that sort. This ability to extract itself from contractual relations is the principal line of flight for such multinational companies. (It is not just Nike, of course; other global brand manufacturers do the same thing.)

The *franchise model* is used successfully by many food production companies like McDonald's. There is no need for the company to own and operate the restaurants that bear its name. Instead, the chain enters into contracts with the people who operate its restaurants and the chain provides them with the ingredients, the corporate furniture, and so on, but the risk of operating each restaurant is transferred to its individual operator. At the macro-level there is also the question of sourcing meat and other ingredients at advantageous prices. Businesses, of course, have incidental effects on the natural and human worlds. For example, the cement industry has admitted that it contributes something like 5 per cent of global carbon dioxide emissions. There was never a better argument for low-rise bamboo buildings.

This is a systemic issue. As Simone de Beauvoir (1962, p 19) observed about French involvement in Algeria, after a while one must recognise that it is not enough to talk of 'abuses' as though they were one-off events, but rather one must realise that this is part of a system. To borrow from Berle and Means, this is a 'corporate system' that is both formalised and legitimised by company law.

What responsibility do shareholders bear?

So, we have to return to the question we asked in Chapter 2: what are the obligations borne by the company? The answer suggested by the House of Lords in *Salomon v A Salomon & Co Ltd* (1897), contrary to the High Court and Court of Appeal, was none. Consider the vast environmental devastation that was caused by the Exxon Valdez when it shed oil across virgin territory in Alaska. There is one way of creating company law that would say that, to force shareholders to control the actions of the company in which they have invested, the shareholders should be personally liable for all of the defaults of the company. The way we have created our company law, of course, is to provide limited liability for shareholders.

Alternatively, there is a third way. We could maintain the principle of limited liability but make exceptions in cases of environmental catastrophe, death or human rights violations, in which shareholders would be personally liable for the company's defaults. The counter-arguments are all based on profitability for shareholders as investors and on the supposed meltdown of our economies if investors stopped investing because they could not acquire limited liability. So, we are left with a choice between a planet which is environmentally secure, in which human rights abuses have remedies, and in which capital is bound by ethical considerations; or a world in which the rich can keep their riches at the cost of the environment and social injustice, and in which the absence of properly funded welfare states is salved by pension funds drawing large profits from liability-free investment.

The sociopathology of the company

Sigmund Freud's *Civilisation and its Discontents* (1929) postulated the idea that human beings are essentially animals with base desires who are required by society to rein in those base desires, and who consequently become torn between instinct and the requirements of social propriety. The lure of profit and the urge to behave in socially responsible ways can be understood as being part of the same division but transferred to the corporate context. In his book *The Corporation* (2004), Prof Joel Bakin suggests that if there was an individual who focused solely on profit and refused to consider the needs of any other person or any other motivation, that person would be considered to be a sociopath. A sociopath is a person with no moral conscience and no concern for other people. An ordinary company is (in law) a person intended to generate profits for its shareholders, unless the articles of association provide to the contrary, owing no company law duties to employees or to creditors. Even the enlightened new s 172 CA 2006 only requires directors to take employees, creditors and the community into account, but does not give them rights to sue the company. Yet this is exactly how we expect companies to behave on the predominant free market model (as discussed below), seeking only profit to the exclusion of any other activity.

Freud identified a large amount of human pathology in the need to conform to social norms and to act against basic instincts. This bled directly into commercial life, though, when public relations and advertising was developed in its modern form in the USA by Freud's son-in-law, Edward Bernays, as he discussed in his book *Propaganda* (Bernays 1928). After the 1939–45 war, the clinical observation of returning war veterans who described their sexual desires and mixed feelings at returning to ordinary society after their gruelling wartime experiences led literally to the belief among advertisers that 'sex sells'. And yet, far from extending the human conscience into the company, companies have come to exploit human psychology as though it were a weakness through which products can be sold and out of which profits can be made. What is more, as we discussed in Chapter 1, company law developed out of trusts law (which is of course based on conscience) but managed to leave the conscience behind as it did so.

So, what does this mean for company law? As Bakin has argued, the purpose of a company within the UK–US company law model is solely to generate profits without thinking about anything else (Bakin 2004). Therefore, there does appear to be a moral vacuum at the heart of the ideology that drives the current capitalist system in which we are all labouring. This is a conceptualisation that is perhaps redolent of Vaughan Williams J in *Broderip v Salomon* (discussed in Chapter 2) considering the modern use of companies to be a misuse of what Parliament intended. Bakin characterises the roles of the corporation in US corporate law as being 'a legally designated "person" designed to valorize self-interest and invalidate moral concern', in that company law encourages directors and shareholders to care only about their own profits at the expense of any larger, moral concerns.

Why do companies indulge in good works?

One of the most acute observers of the limitations of CSR theory and a cynic about corporate giving is Professor Robert Reich in his book *Supercapitalism* (2008). Reich is concerned that the real worries with the behaviour of companies are being subsumed into a form of CSR thinking which is popular at business schools and which has been purportedly embraced by companies who are pretending that a little charitable giving constitutes a new, enlightened thinking. As Reich put it: ' "Corporate social responsibility" has become the supposed answer to the paradox of democratic capitalism', although ultimately the company is all about profit, it is all about 'the bottom line'. Instead, what purports to be social responsibility is in fact simply public relations and an exercise in cutting costs (so as to increase profitability) while spinning for PR purposes a line that cost-cutting is really an exercise in good public works. For example, as Reich points out, Dow Chemical reduced its carbon emissions so as to cut its costs but trumpeted the effort as being instead an exercise in social responsibility.

One of the most evident examples of a British company purporting to act in a socially responsible manner is the oil company BP, which has been by definition a huge contributor to carbon dioxide emissions over the decades by producing petrol. BP announced that it was going to spend a large amount of money on alternative fuels as part of a rebranding exercise in which its corporate logo became a sort of stylised flower head. However, when one considers the numbers involved, it is not so clear that the company actually has an overriding commitment to alternative fuels in that in one financial year it spent about US$800 million on alternative fuels, while making profits of US$20 billion with total expenses on its business of US$14 billion. Therefore, spending on alternative fuels was a fraction of its total cost base in relation to oil production. In Reich's view, 'global supercapitalism does not permit acts of corporate virtue that erode the bottom line' because 'shareholders do not entrust their money to corporate executives for them to give it away, unless the return is greater'.

So, Reich argues that CSR has become a feint in which large companies (the so-called 'supercapitalists') have noticed increasing concern about their impact on the planet and on society, and so have used corporate images, marketing and so forth to position themselves as caring members of society. It should be recalled that many multinational companies behave very differently in different jurisdictions. For example, oil companies are frequently less concerned with employee contentment in the countries that produce the oil than in countries where the petrol is sold. Armed guards are retained where the oil is pumped and refined, but smiling adverts are run where the petrol is consumed in its largest amounts.

The global meso-economic context

Supercapitalism is predicated on a simple imbalance in world trade: in the 'developed world' incomes are high, whereas in the 'developing world' incomes are low. Consequently, supercapitalists are able to manufacture their goods at low cost in countries where incomes are low and then sell them at a high price in countries where incomes are high. This system is clearly based entirely on inequalities in wealth around the world. So two of the principal issues for company law policy in the 21st century are the need to consider how responsibility is removed from investors by the *Salomon* principle and also the level of regulation that is required of trading companies in different contexts. At present, removing the regulatory and moral elements from our company law is harming our economy as well as challenging our collective ethics.

On Marx's model ordinary workers are alienated from their labour because their labour is bought by the owners of the means of production, but they have no direct stake in its quality or its profits. This is most evident in modern sweatshops in Asia producing textiles and footwear, where the workers are paid a pittance while the owners of the branded products charge extraordinary amounts for the products. On both models, something pathological is caused by the organisation of capitalist society. CSR, essentially, tries to put this phenomenon into a broader context than company law or an accounting focus on pure profit will allow.

Moving on ...

Currently, company law is formally unconcerned with companies' treatment of its employees and the rest of society. The closest company law has come to accommodating these issues is in the reorganisation of the duties of directors in s 172 CA 2006, as discussed in Chapter 7, requiring the directors to promote the success of the company and in so doing to consider the position of employees and the community more broadly. What remains to be seen, as considered in Chapter 7, is how the courts will conceive of this duty and whether it develops a meaningful extension of company law into something that could reasonably be thought of as constituting social responsibility. This idea is considered in the next chapter.

Chapter 15

Thinking about company law

Introduction

Company law as an ideology

The ideological place of company law

Company law is part of an ideology. It is part of an ideology about the way in which our society should function. Ever since the decision in *Salomon v A Salomon & Co Ltd* (1897), it has become a part of our national culture that investment and economic growth were more important than an old morality about an individual being responsible for their own actions and their own debts. The House of Lords gave legitimacy to the idea that businesspeople could protect themselves from loss (and thus impose the risk of loss on everyone they dealt with) by using a company. Companies proliferated wildly as a result. The effect was twofold. First, companies became the most significant economic actors in our economies and acquired a personality in the minds of ordinary people as well as in our company law. Second, a sense of unreality took hold in that we all accepted that these abstract entities controlled so much of our lives. We accepted that these intangible legal persons should be treated as though they were people because of the status that company law gave them.

To illustrate the way in which the unreal became real, let us take the example of the large public companies that operate supermarkets. In the ordinary course of things a supermarket chain was identified with the people who worked in our local branch and in the logos that flew on advertisements and over their shops, but there should also have been a sense of an impersonal organisation providing most of us with the food we ate. It is only in the 21st century that popular culture has begun to think about the world behind those logos: the farmers who produce the food, the additives which go into it, the animals which are hidden in this corporate food chain. The power of companies is exercised in another layer of unreality and of mere words: in our company law, in the accountancy jargon, and in the financial prospectuses that lure in investors. The needs of intangible persons (that is, companies) began to dominate the debate about what constitutes a successful economy and what our political priorities should be. We treat the 'needs of business' in discussions of public policy with at least the same reverence as the needs of human children living in poverty. In those senses, company law directly affected the ideology of capitalism and business that took hold in this country and around the Western world.

So it is true to say that our company law should not be thought of simply as an arid zone intended solely for legal professionals. Instead, the content of our company law directly affects our public life and our culture. If the law stands for our moral and cultural understanding of how our society operates (something which we accept almost automatically when we talk of criminal law), then our company law stands for the moral and cultural way in which we see our economic life. This chapter considers where we stand early in the 21st century in that understanding of our commercial life, and what the possible futures are for company law within the economic system.

Company law and capitalism

It seems difficult to imagine now, but there was a time in the late 20th century when it was not entirely clear whether a country such as the UK with a welfare state would be considered to be capitalist or to be socialist. Of course, much of this is a matter of perspective. For a US congressman from the Deep South, the very idea of public healthcare like the National Health Service in the UK is purely socialist. Yet, in truth, English company law from 1897 onwards was always concerned to make capitalism possible. It regulated the interaction of investors and managers within a business, and hermetically sealed the business of the company off from the outside world so that the company could focus on internal discussions about the corporate constitution, the wishes of the majority shareholders under, for example, the *Foss v Harbottle* rule. Therefore, the post-*Salomon* form of company law stood for a particular brand of capitalism in which the British 'workshop of the world', as we described ourselves in the old history books, was peopled by humans and companies who worked hard, lived within their means and had a real place in the world.

However, since free market economics took hold of the levers of public policy in the late 1980s, we have moved into a world with a new kind of capitalism. In the UK and in the USA in particular, banking and financial services became the most important businesses because they promised growth and investment without the damaging inflation of previous decades. What this movement signified was a movement away from real, physical production into an abstract world of finance as the driver of economic policy. Developed nations saw their production shift offshore to cheaper producers, and a world of financial services took its place, to the extent that an estimated 25 per cent of the tax take in the UK was thought to come from the financial services sector directly or indirectly (although others have put this figure at something more like 12 per cent). Either way, the UK economy had become orientated around the financial sector to a remarkable extent.

The global financial crisis that began in 2007 possibly signalled the beginning of the end for this kind of capitalism. It is estimated that it will cost the world economy around US$12 trillion before it plays out entirely. After all, capitalism is just another '-ism', just another ideology. At its root 'capitalism' refers to an ideological preference for the means of production and economic power to be held in private hands by capitalists, as opposed to being held in public hands in a genuinely socialist system (Van Parijs 1993). (Most so-called socialist parties are actually social democratic parties that accept the operation of economic life primarily by entities in private hands but with a system of regulation to restrain or guide their activities and a welfare state to provide a safety net for the most vulnerable. In truth, most right-wing parties agree with much of that in practice, but they quibble about the amount of State involvement that is appropriate in all of these activities.) The new capitalism sought to reduce the level of State participation. It was an ideology in which the development of financial products (such as derivatives) was expected to remove risk, in which it was possible to have constant economic growth without inflation, and in which deregulation of economic activity would provide the liberty necessary for economic actors to provide constant growth. All of these phenomena were contrary to the received economic wisdom of the past.

And none of it was true. The constant growth which boomed in the developed world during the 1990s and up to 2007 was predicated in large part on shifting away from manufacturing to selling financial services, on a perpetual boom in consumer spending, and on ever-rising house prices which made ordinary people feel wealthy. In fact, it was the acceptance of low, poverty wages in many of the developing countries which were producing the goods that were consumed so avidly in the developed world which made the consumer boom possible: low wages for workers in the developing world meant low prices in the developed world, which meant high rates of consumption. A huge amount of this consumption was fuelled by consumer debt, and not by cash. The boom in the USA in particular was fuelled by Chinese investment in the USA as China sought to prevent inflation at home by spending its surplus cash overseas. The inflows of Chinese investment created a bubble of money that was available to borrow at low rates of interest (because it was so plentiful). Remarkably, it was the communist Chinese government that was funding American capitalism. When the US economy began to turn downwards under George W Bush and eventually crashed, the reverberations were felt around the world. The clearest example of the effects of this new world order was felt in the global financial crisis that began in 2007.

The causes and the legacy of the global financial crisis

An interconnected world economy, in which large financial institutions invested in overseas markets and in particular used financial derivatives to speculate on market movements, meant that a serious collapse in one financial market would pull down all of the others. (See Hudson 2009b, p 28 on systemic risk.) So when the US sub-prime mortgage market began to collapse in 2007, there were serious ramifications for the entire financial system. The securitisation products (in particular collateralised debt obligations (CDOs)) that had enabled mortgage lenders to shift the risk of borrowers not repaying their mortgages onto third party investors meant that all of those investors suffered losses when the crashing economy caused those people to fail to make their repayments. Worse than that, however, was the amplification of that crisis by the CDS (credit default swap) market in which speculators were taking positions on the CDOs collapsing because the borrowers would fail to make repayments. The collapse in the CDS market taken together with the collapse in the US real estate market was what caused the investment bank Lehman Brothers to fail: the largest corporate insolvency in US history. And when Lehman Brothers failed, the entire banking system stopped working because no banks would risk lending money to other banks. This meant the collapse of many, many ordinary businesses because they could not get funding. (See Hudson 2009b, Chapter 32 on the financial crisis.)

What this signalled was the failure of this particular capitalist system. It was a slap in the face to the free market theorists because governments around the world were required to intervene to invest directly in banks which otherwise were insolvent, to make borrowing facilities available for the other banks to provide them with short-term money to get through the difficult patch, and to bail out large non-financial companies too. It became clear that the market could not be trusted to act in its own (let alone anyone else's) best interests, and that there would need to be regulation. It also became clear that the economy could not function without direct intervention from governments. The free market hypothesis was simply wrong.

So, what does this mean for company law? It means that we cannot accept the free market hypothesis that has underpinned so much company law scholarship for so long unthinkingly. We can no longer accept at face value the ideology about the need for management to be free to act in the best interests of the company (and thus the broader economy) and for companies to be free from regulation. The development of the directors' duties in the Companies Act 2006 (as discussed in Chapter 7) demonstrated tremendous foresight in this area.

It has been argued by some that the financial crisis that began in 2007 was a 'perfect storm': that is, that it was a freak occurrence resulting from entirely random factors coinciding. This is nonsense. Others and I had been arguing for some time that that sort of crash was inevitable (for example, Hudson 1996, p 228; Hudson 1999, p 235). More significantly, there are regular crises in capitalist systems, so this was not a one-off collapse. Indeed, Friedman himself identified crises with the development of capitalism (Friedman 1962), as though a crisis clears out the dead wood to allow new growth. In Chapter 9 we considered the effects of deregulation on the Enron and WorldCom collapses – two of the largest collapses before Lehman Brothers. We might think that this was just coincidence. But that overlooks the crashes caused by the hedge fund Long-Term Capital Markets, the Japanese economy, the Russian banking moratorium, the rapidly increasing price of oil just in the 1990s, as well as failures in several financial institutions before the clear-out of 2008. Then in 2010 we saw financial markets driving nation states to the edge of insolvency – in particular Ireland and Greece, but titans like the USA and the UK also had to enter dangerous budgetary waters to cope with deficits of a size not seen since the Second World War. Ironically, the same banks who had relied on government bail-out money only months before were now fixing the price of government borrowing so high that some of those countries required bail-outs, as well as to spend taxpayer money on increasing interest rate costs. This must make us wonder whether this version of capitalism is working if the banks can cause such mayhem and soak up so much public money, while leaving the costs to be borne by ordinary people, before returning to profit one year later while driving nation states further into insolvency. The advent of democracy movements in the Middle East in early 2011 have caused further turmoil in energy markets: oddly, the price of oil keeps hitting new highs, but after a crisis it never seems to return to old lows.

In truth, this capitalist system lurches from one crisis to another in this way. What is happening is that the sharp reverses which are felt in ordinary commercial markets are amplified on the national stage and at the level of the global financial system. That is why allowing the capitalist system to replace the State is so dangerous, because the stakes at the national level are much higher than short-term competition in a commercial marketplace.

In this chapter we shall begin to unpack some of these arguments and see how the ideology of modern company law sits with the world as we experience it all around us. In essence, the effect of a company law and a capitalist ideology in which the internal profitability of the business is all that matters cannot be sustained. In a changed world after the financial crisis and an increasing realisation among consumers in the developed world that the old ways are no longer fit for purpose, we need to imagine how things might look differently in the future.

Capitalism old and new, and newer still

In an excellent book on company law, *Company Law and Capitalism* (1972), Professor Tom Hadden considered, inter alia, the contextual way in which company law functioned at the intersection of the desires of capital and the needs of workers. That was a time in which the 'factory floor' and 'the boardroom' were considered to be the places in which disputes were acted out. Indeed, the labour unrest in the UK in the 1970s (culminating in the 'Winter of Discontent' which saw Margaret Thatcher come to power in 1979) appears to be a relic from another age, which simply demonstrates how quickly new ideologies can take root. Indeed, it is possible to write a book on company law like this one without needing to talk about employment law or environmental law, as though company law could operate in a sealed box which is insulated from the concerns of workers, of society more generally and of the broader environment. Therefore, it is a source of optimism, perhaps, that even though the manufacturing base and heavy industry in the UK (such as coal mining, shipbuilding, car construction) have withered away, nevertheless the growth of corporate social responsibility, corporate governance and the new code on directors' duties have begun to reintroduce the concerns of people outside the triangle of company, directors and shareholders to debates about company law.

What remains to be seen is whether or not we will move into a new form of company law in which the judges choose to give effect to these directors' duties, for example, in a progressive way, or whether the judges will choose to interpret the need to take into account the needs of society more broadly as being subsidiary to the best financial interests of the company. In the modern world we would want the old permission for directors to buy cakes and ale occasionally for the workers to expand into an understanding that profitable production does not mean squeezing out the last ounce of profit at the expense of pollution or harm to others. A corporate model that is based entirely on earning profit at the expense of anything else (Friedman 1962) necessarily requires the managers to ignore the marginal cost of making profit. Instead, a sustainable model of company law requires that a reasonable expectation of profit should now be mixed with an obligation to do no harm to the environment, no harm to the interests of workers, nor any harm to other stakeholders in the business. (The idea of stakeholding in this sense is considered below.)

What also remains to be seen is whether we will move into a new form of capitalism. In the wake of the financial crisis, and with the growth of the environmental and human rights movements around the world, it may be that we will move into a more enlightened form of capitalism in which the pure profit motive is tempered by a realisation that as a species human beings need to take better care or their planet and of each other. To do this, the type of capitalism which has proved itself to be so untrustworthy at the start of the 21st century will need to be reined in by governmental and regulatory control, as well as by a different ideology among consumers who are concerned about where and how the goods which they buy and the services they use have been produced.

One-dimensional legal people

The social theorist Herbert Marcuse wrote a book called *The One-Dimensional Man* (1964), in which he identified the way in which people were locked into a simple obedience to a single set of beliefs and behaviours, which meant that they appeared to live only in one dimension without the possibility of individuality or genuinely distinct personality at a psychological level. It could be said that our company law has locked the company into a single set of beliefs and behaviours that mean that they appear to operate only in one dimension without the possibility of a genuinely distinct legal personality at anything other than the legal level. (Do you see what I did there?) While our understanding of the human condition may have improved (with the recognition of human rights law and a general improvement of the quality of living in this jurisdiction in my lifetime) or worsened (with the increasing atomisation and trivialisation of our private and public life), it does seem as though our understanding of company law is locked into a one-dimensional framework. One of the dimensions which has been missing, as suggested in the last chapter, is a moral understanding of the impact of its actions on other people – something which was positively encouraged by free market economists like Friedman.

The company law of the 20th century developed a one-dimensional company in which the overwhelming objectives of the directors were to deliver a financially successful company for the shareholders, and the shareholders had a legal power to dismiss the directors if they did not. By the same token, commercial practice also developed an idea that businesses must constantly seek profit growth year-on-year for their investors. This sort of pressure drove companies like Enron to commit fraud rather than admit to losses, and it drove most other public companies into skewing their business objectives so as to meet these sorts of growth expectations among investors. As a result, companies in this jurisdiction were prevented from developing other dimensions that would require them as part of company law to operate best practice in their corporate governance and to take into account the wishes and needs of employees within the company. In the next section we shall consider some of the alternative conceptions of the company and how they might usefully interact with company law in the future.

Some of the shibboleths of company law thinking

The danger of thinking of the company as being merely a nexus of contracts

As discussed in Chapter 1, the English joint stock company began life historically as a contract: that is, as a partnership of traders with their joint stock being held on trust for them by the managers of their business. The separate personality of the company is, in consequence, a judge-made recognition of commercial practice that breathed life into the company itself and allowed the shareholders to slink into the shadows with their limited liability. When asking the question in the 21st century 'what is the nature of a company?', it is therefore frequently said that the company is best described in legal terms as a 'nexus of contracts'. Professor Ronald Coase was the principal mover of this theory in which the company was seen as a sort of black box in which all of the various transactions which take place in a business are simplified by recognising the personality of the company (what Coase, an economist, refers to as 'the firm'), which removes the need to complicate that process (Coase 1937). This analysis has some attraction purely on its face as a means of describing the status quo in relation to the organisation of rights in companies; it also reflects a historical truth about the genesis of those legal structures that gave birth to the modern company post-*Salomon*.

Yet, the ideology at work here is the depersonification of the company by reducing it to a mere investment. Professor Paddy Ireland has described this process of reducing the responsibilities of the shareholders to nothing as a result

of thinking of shareholders being merely investors as being an attempt to 'legitimate the tenuous claims of vestigial, rentier shareholders to a significant part of
the social product' (Ireland 1999). Thus, without bearing personal responsibility
for a company's losses, the shareholders take the profits in the form of dividends.
Their interaction with the company becomes a mere investment contract, just as
the contract itself is thought of as being a nexus of contracts to simplify trade.
What has happened is that there has been a sort of disappearing trick in which
none of the human beings bear any theoretical responsibility for the acts of the
company because its legal personality provides them with shadows in which to
hide. Consequently, to reduce the company to mere contracts removes the fiduciary context, the possibility of 'ownership', and the possibility of conceiving of
the company as a community of investors, managers and employees.

Among the labour law commentators, Professor Hugh Collins has expressed
his concern that conceiving of the company only in terms of contract masks any
possibility of recognising the claims which employees and others ought to have in
relation to the company, primarily because they are persons outside the ambit and
privity of the contract that is expressed in the articles of association between the
shareholders and the company, as embodied now in s 33 CA 2006 (Collins 1993).
To restrict the discussion to a form of investment contract between investor and
investment entity is to ignore the far more significant social and economic role
that the company plays in society more broadly and to relieve the company of
any responsibility for its impact on the broader community. The question that is
considered in the following sections is how to conceive of the company other than
as being simply a contract of investment. As cases like *Mercy Docks and Harbour
Board v Coggins and Griffith* (1947) illustrate, the employment contract contains
direct consequences for employees that lack any of the nuance, flexibility or
remedies associated with commercial contracts. Lord Simmons pointed out that
an attempt by an employee to renegotiate his terms of employment, while 'sturdy',
would lead to dismissal; whereas a commercial transaction often permits renego-
tiation and adjustment (Masten 1993, p 196). Thinking of a company as being
about contract forgets that the history of employment contracts has always been
different from that of commercial contracts [.] The contract analogy is simply
misleading as a result.

The alienation of the humans who work in companies

One of the longest-lasting Marxian concepts has been the alienation of the worker
from her own skills over time as capitalism required the worker to work for a
wage which was unrelated to the value of the goods produced (as was discussed
in the previous chapter). A new feudalism was created by the industrial revolution
and the development of the company in which agrarian labourers growing subsist-
ence food or generating produce for sale at market were removed from contact
with the land and became workers in the industrial towns. E P Thompson (1963)
identified a single working class that grew out of this alienation from its own

work and production. Hannah Arendt (1998) drew a distinction between various kinds of worker: those who were obliged simply to 'work' in menial tasks which imprisoned them and those who were comparatively free to engage in their skilled 'labour'.

In the jurisprudence today there is still a need for a strong overlap between the law relating to the employment relationship, the governance of the company, the rights of members and the obligations of companies to third persons to understand in the round how companies interact with the entire economy. The company altered the straightforward bond between something English law used to refer to as 'master' and 'servant', individual and State, by interposing its own distinct personality. From this seed, companies grew to dominate our economic life: a little like a young plant growing between the cracks in the paving flags and forcing them apart.

The company is the master now. Whether in the form of financial institutions, industrial concerns or co-operative ventures; whether research-based institutions, predatory capitalists or global media undertakings; whether eco-friendly or politically incorrect – the corporation is king in this new world. This power is maintained in capitalist societies in Gramsci's terms through hegemony (dominating political debate and dominating so much of our culture) and thus through ideology. In Chomsky's terms, this is achieved through the maintenance of necessary illusions by corporations over individual citizens (Chomsky 1988 and 1989). The manufacture of consent is achieved by the transmission of messages into society – through advertising, through political debate – that the current system offers us security, prosperity and a bright future. The necessary illusions include an ideology of growth-bringing companies driving the economy forwards. A large part of that illusion is the idea that companies are governed by law and that the danger of corporate scandals will be contained by non-legal corporate governance standards.

What this ideology attempts to mask is the presence of risk in our market economies, as exemplified by the financial crisis. The citizen is reliant on the performance of the interconnected, global economy and of individual companies to provide her with employment. To convince the citizenry that it is sufficiently endowed with rights and access to riches has required the generation of complex social myths through the ideological canon of media, education and State institutions to promote capitalist values of investment and economic efficiency. Yet those same theorists, like Friedman, accept that crisis and the risk of loss is necessarily a part of the system. A very significant part of this form of capitalism is the company as an individual actor making ostensibly rational and efficient decisions: not as an expression of the solidarity of its membership of society, but rather as a pseudo-sentient creature able to speak without a face.

The company as a community: a challenge to company law

In this book we have presented the straightforward company law analysis of the shareholder as being the only possible owner of the company's property and the only person entitled to share in its profits. As we have noted, company law tends to overlook the employment law context in which employees also seek to participate in the profits of the company because company law conceives of the employees as being creditors of the company and not participants in its affairs in a legal sense. This is not the way in which economists or other social theorists think about companies. They tend to think about the employees and the managers as being motivated by the success of the company and by the money that they are paid by the company. Lawyers focus on legal rights and obligations of the parties, and they think of employees and creditors as having no rights other than those under contract law. By contrast, other social sciences think about companies as having a broader range of 'stakeholders'.

Stakeholding itself can be seen as combining a concern for actors like creditors and consumers with a need to see the company as a broader coalition of interests than simply the economic expectations of shareholders. Conceiving of different actors as having a stake in the company requires us to think of the company as being part of a broader community and, in the mind of some commentators, makes the company a form of community in itself. One example is the influential management theorist Charles Handy. His communitarian thinking advocates a recognition of the company's separate personality so that the company is conceived of as a 'public body' (that is, a body with a role in the outside world) which interacts with the community at large while itself constituting a community peopled by managers, employees, shareholders and others. As Handy puts it:

> We should think of a business not as a piece of property owned by someone but as a living community in which all the members have rights and whose purpose is continued existence. Companies are communities of people united by common aspirations rather than a bundle of assets owned by shareholders. . . . In the increasingly knowledge-based company, the company's real assets lie in the brains of its employees. It is all the more important that, as key members of the company community, they should also have rights. (Handy 1996, p 28)

This is a structure which Handy himself dubs a 'stakeholder company', which relies on the concept of a community of interests in which everyone works for the success of the company if they feel that they are a part of it, and not as if they are just another pair of hands toiling for the profit of others. The idea of a 'community' here necessarily begs a number of questions as to the extent of that community, the extent to which its members can be said to have sufficient ties to hold them together in a socially meaningful way, and then the nature of their interaction. To talk of the company-as-community being a unit made up of all who fall into defined categories (shareholder, director, employee, creditor and so forth) is merely to compile an attendance register of all of the people who come together in the activities of a company. What is important is whether their interaction is of a quality that necessarily makes them feel emotionally that they are a community with a common goal: this is something which company law prevents by excluding employees, creditors and others from having a legal voice in the activities of the company.

The only crack in this wall is the requirement that directors consider the position of employees, creditors and society more generally further to s 172 CA 2006. For the CSR movement and those, like Handy, who would like companies to become more human places to work, the duty on directors to promote the success of the company, as discussed in Chapter 7, could lead to a transformation of the institution if directors are not able simply to make a show of considering these issues before putting them to one side. This does require traditionally conservative English judges in the Chancery Division and in the Commercial Court to think more creatively about these new concepts which Parliament has created than the narrow-mindedness of decisions like that of Lord Hoffmann in *O'Neill v Phillips* would allow.

At present these concepts suffer from a level of underconceptualisation. They beg the question as to how this community would work and what form of rights each of the various actors would have and how disputes between their necessarily competing claims over the assets of the company would be reconciled (Ireland 1996, p 301). At one level, its most attractive feature is a determination to humanise the company by reconceiving of it as an expression of the collective undertaking of all of those individuals who interact in that space. The most useful application of the community ideal is perhaps as a metaphor for the need to understand the company as a space within which activities take place, lives are lived and sectional interests fought out.

The company as a cypher, and its moral consequences

Woody Allen has a stand-up routine in which the conceit is that he has incorporated himself as a company, needing to use some of his relatives as directors to satisfy company law, only for those relatives to drive him out of his own company by taking over the board of directors (Allen 1999). This is of course possible if those directors can outvote the individual who created the company. It is also bizarre. Think of it this way. The company is equivalent to your imaginary friend. You conjure it up to help you do things but it has no tangible existence; just as the company does not actually exist, except for the fact that you and the law pretend that it does. You are able to do things with your company just like you could with your imaginary friend: that means you actually have to do everything yourself but you tell everyone else that it was your imaginary friend (that is, the company) that did it. The difference is that under company law the other directors or any other shareholders can take control and throw you out, whereas other people cannot take over your imaginary friend and refuse to let you see it again. The difficulty with legal fictions like a company is that there are always circumstances in which the logic begins to strain.

There is an interesting science fiction novel, *The Unincorporated Man*, by Kollin and Kollin (2009), which imagines a future in which all human beings are required by law to organise themselves as corporations from birth, so that they are owned by other people as their shareholders. The hero is a man who has been held in suspended animation for centuries so that he was never incorporated. He is the hero (in a series of novels) in that he is free from the burden of satisfying the corporations who own shares in all of the other people and which therefore control their lives directly. The unincorporated man is free and so becomes a beacon for other people who want to break free themselves. The novel is based on an idea from the free market economist Milton Friedman's work, in which he suggested that people in the future should incorporate themselves so that they could sell shares to investors, which would entitle those investors to a share in their future earnings, with the result that people would be able to raise money for their educations and so forth (Friedman 1962). (Of course, this is organised at the moment on the basis of debt, whereby people borrow money instead of issuing shares.) All of this illustrates how *the very idea of the company* has possibly still not finished developing. Instead, it is a creature that human beings can continue to sculpt for a wide variety of uses, just as Dr Frankenstein did with his monster.

Salomon old and new: a tale of two capitalisms

At the end of Chapter 2 I mentioned the Salomon corporation that exists today making shoes, among other things. There is no connection between the Salomon & Co Ltd in the 1890s making boots for the army, and the 21st-century Salomon companies. Nevertheless, comparing the two businesses is illuminating. The establishment of A Salomon & Co Ltd is an example of extraordinary legal and commercial ingenuity. It also shows us how laws can reap unintended consequences: a statute intended to ensure probity in the boardroom by requiring a minimum number of members became instead a means of organising your affairs to avoid personal bankruptcy.

By contrast, the Salomon training shoe and ski-wear company that we know today began as a small-scale manufacturer of ski edges in Annecy, France, in 1947 (so just after the end of the 1939–45 war). It was a family company, under the leadership of George Salomon. The company website even has a photograph of its original unprepossessing workshop (www.amersports.com – follow the prompts to the corporate history). The development of a new toe piece for ski boots in 1955, and its adoption by a famous skier in 1961, seems to have changed the business into a high-profile ski manufacturer. It was not until 1992 that the company broke out of the ski equipment business into manufacturing hiking boots and so on. Before then, the business had been founded on success in competitive skiing competition and the sale of branded ski equipment. That is, until Adidas bought the business in 1997 to form a new company: adidas-Salomon SA. (Note the use of the lower case 'a' in adidas, even though my spellchecker wants to capitalise it. The 'SA' is the French term for a limited company: Société Anonyme, or 'anonymous association'. I like that idea of an organisation without a face – it sums up a company perfectly.)

This new company began to focus more and more on footwear and in essence combined adidas's international brand name in casual training shoes with the newly re-energised Salomon brand for healthy outdoor activities using technically advanced footwear. While Salomon continued to make competitive ski and winter sports clothes and equipment of all kinds, the brand image that is most visible now on the high street is in hiking equipment (buying into the healthy-living boom) and in leisurewear. The brand is always based on a technological image ('motomesh', 'softshell', '3D') and a link to cool activities (pun intended) like snowboarding, skiing and even surfing. In 2005 the business was sold to Amer Sports.

As an illustration of the modern corporate business model, we can see how a global goliath acquired a brand to expand its own product range into a new market, and linked the brand to sporting success and to cool leisure pastimes. Every photograph involving the brand seems to be on top of a mountain looking over a sunny winter view, or is of a snowboarder making an impossible jump in the snow. The suggestion of cutting edge technology is always close at hand. In line with the discussion in the previous chapter of identifying corporate 'values' and linking them to social causes, in 2005 Salomon made a '365 day commitment to women' with the 'Women will' and 'Live your dream' promotions. This could be seen on its face as promoting the rights of women to mountainside leisure activities, or simply as a marketing ploy to reposition the brand by selling products to women and not just to men. As a mission statement, the company identifies its values as being 'Authentic: Commited: Progressive'. (So committed in fact that it spells 'commited' wrongly.) These values are of course illustrated with aspirational photographs, rather than words or proof. In themselves, they mean nothing. What, after all, is a 'committed' hiking boot? Or, more to the point, what would an 'uncommitted' hiking boot be like? But it is de rigueur for a global brand to be identified with the sorts of warm words that they think will appeal to their key demographics. As consumers we know this and so we tend to cleave towards slogans and images that we think are directed at our own aspirations.

Like all global brands, there also needs to be product growth. So, not content with continuing to manufacture leading ski products, or latterly hiking boots or high-concept cross-country running shoes, they now also sell 'apparell' (sic). Again, this apparel is aimed at outdoor sports. Salomon are beginning to use urban backdrops (particularly concrete walls) for some products, instead of mountainsides, which signals a move towards the huge numbers of people who wear running shoes to slouch to the corner shop or to shuffle around the workplace.

The corporate advertising speak has taken over (but spellchecking has not) as this company moves into a different form of capitalism from that which prompted George Salomon to open a workshop in Annecy in 1947, let alone that which prompted Aron Salomon to open a boot factory in Whitechapel in the late 19th century. Salomon now has 40 distributors selling into 160 countries around the world. It has a corporate vision that is now about selling 'footwear, apparel, equipment and experiences' around the world, even though only three of those things are actually products. Ironically, perhaps, this new form of capitalism began with a bootmaker called Salomon protecting himself against the risks of the future, and it has reached a peak where another manufacturer of 'footwear' and 'experiences' is marketing a Salomon brand in almost every country on the planet while its products are made in factories in China (in the case of my trainers).

Perhaps boots that were made for use by the British army at the high-water mark of the British Empire in the 19th century would have travelled to many of the countries on the planet, but they were made in the East End of London. In the 21st century there is a different sort of imperialism: that of the huge corporations with their global brands and their direct political influence in developing countries. It is not an imperialism built simply on armies and occupation. It is more subtle than that. It is a 'soft power' of investment and promising economic growth in countries like Indonesia, which grants such power to global brands within designated development zones away from the inconveniences of minimum wage and health and safety legislation in the countries where those branded companies have their headquarters and their principal markets.

The boom years of the 1990s and the early 21st century were built on the low wages and poor working conditions in the developing world generating cheap goods and services for the high-wage economies of the developed world. (All that is 'developed' in this context, of course, is the capitalist infrastructure.) Cheap money in the USA was funded by Chinese investment of its spare cash into the USA. China was effectively subsidising the purchasers of the goods it manufactured. The developed world shifted from production to a period of consumption. As the developing world demands democracy and a better standard of living, the fragility of this model will begin to show through.

Apparently, the only things that are to be publicly funded in developed countries in today's world are failed investment banks and the military. The old hippies asked to 'make love, not war'. We make war and we make money. Love has to be paid for privately (if you see what I mean).

Conclusion

My own view is that there is a need for greater transparency in our social life. This transparency requires that companies not be entitled to treat themselves as facades that can close off the moral responsibility of those within the company. Rather, in a society that has become hyper-complex, it is important that private reservoirs of power are subject to the public democratic gaze. As a growing proportion of the population becomes socially excluded, the private commercial corporation is an ever-increasing locus of power and an ever-growing cause of that exclusion. The welfare state is rapidly ceding ground to corporations in relation to pension provision, employment patterns and other services (through the contracting-out of those services and through the discredited Public Finance Initiative and Public–Private Partnership schemes).

In this context, the company must be transparent to the democratic gaze of the population that is affected by it. That means that the human actors who populate and drive those companies must be morally and practicably accountable for those things that are done in the name of the company. In consequence, this requires a radical reappraisal of the separate personality of companies and the possibility of the direct personal liability of shareholders. While the prevailing ideology of the free market capitalists tells us that this would be disastrous, perhaps we should remember the essential truth about law, as the French philosopher Michel Foucault (1969) explained it: things are only the way they are because we *say* that they should be this way. Our company law is not something that exists as part of nature. Instead, words put together by Parliament and by the courts brought it into being. Companies only have separate personality because we accept that they do. Under the present state of the law, a change to the decision in *Salomon* would stand much of commercial life on its head by making the stakeholders inside companies at least morally responsible for the things that are done in the company's name.

Much in these last two chapters has cast a shadow over the technical ideas that were considered in the earlier chapters. It is worth remembering that many of the more egregious activities of companies are committed by large multinationals whose human operators have succumbed to a profit-orientated groupthink. However, the company limited by shares is just one among many models in English law for organising social interaction. Nevertheless, it has been spectacularly successful. It is just a model, though. Like Dr Frankenstein contemplating the power in his hands, we could use our powers and our companies for good or ill. At one level, company law does seal shareholders and directors away from responsibility for the actions of their organisations. In particular, it is to be hoped that the new code on directors' duties might begin to change the ideology that underpins our companies. And so, at another level, we should remember that any fault does not rest with the abstract entity that is the company, but rather any fault must rest with the human beings who operate them. After all, nothing is either good or bad; rather, thinking makes it so.

Bibliography

Allen, W, *Woody Allen – the Nightclub Years 1964–68*, available on EMI, 1999.

Arendt, H, *The Human Condition*, 2nd edn, 1998, Chicago Press.

Bakin, J, *The Corporation*, 2004, Free Press.

Balen, M, *A Very English Deceit*, 2002, Fourth Estate.

Bauman, Z, *Liquid Modernity,* 1999, Polity.

Berle, A and Means C, *The Modern Corporation and Private Property*, 1932, Harcourt, Brace & World.

Bernays, E, *Propaganda*, 1928, Liveright.

Bratton, R, (2002) 76 *Tulane Law Review* 1.

Carr, CA, *The Law of Corporations*, 1905, Cambridge University Press.

Chomsky, N, *Manufacturing Consent*, 1988, Pantheon.

Chomsky, N, *Necessary Illusions*, 1989, Pluto Press.

Coase, R, 'The nature of the firm', 1937, reproduced in Williamson, O and Winter, S (eds), *The Nature of the Firm*, 1993, Oxford University Press.

Collins, H, 'Organisational regulation and the limits to contract', in McCahery, J, Picciotto, S and Scott, C, 1993, *Corporate Control and Accountability*, Clarendon Press.

Company Law Review, *Modern Company Law for a Competitive Economy: Developing the Framework* (URN 00/656)

Cooke, CA, *Corporation, Trust and Company*, 1950, Manchester University Press.

Cooper, G, *The Origin of Financial Crises*, 2008, Harriman House.

Cotterrell, R, *The Sociology of Law*, 2nd edn, 1992, Butterworths.

de Beauvoir, S and Halimi, G, *Djamila Boupacha*, trans P Green, 1962, Macmillan.

Foucault, M, *L'Archeologie du savoir*, 1969, Editions Gallimard; translated as *The Archaeology of Knowledge*, 1972, Tavistock.

Fox, L, *Enron: The Rise and Fall*, 2002, Wiley & Sons.

Freud, S, *Civilisation and Its Discontents*, 1929, Hogarth Press.

Friedman, M, *Capitalism and Freedom*, 1962, Chicago University Press.

Fromm, E, *The Fear of Freedom*, 1942, Routledge & Kegan Paul.

Fukuyama, F, *The End of History and the Last Man*, 1993, Penguin.

Fuller, L, *Legal Fictions*, 1967, Stanford University Press.

Galbraith, JK, *The Great Crash*, 1955, Hamish Hamilton.

Galeano, E, *Days and Nights of Love and War*, trans J Brister, 1983, Monthly Review Press.

Hadden, T, *Company Law and Capitalism*, 1972, Weidenfeld & Nicholson.

Handy, C, 'People and Change', in Radice, G (ed), *What Needs to Change?*, 1996, HarperCollins.

Hansmann, H and Kraakman, R, 'The End of History for Corporate Law', http://papers.ssrn.com/paper.taf?abstract_id=204528.

Hayek, F, *The Road to Serfdom*, 1944, Routledge.

Hudson, A, *The Law on Financial Derivatives*, 1st edn, 1996, Sweet & Maxwell; 4th edn, 2006.

Hudson, A, *Modern Financial Techniques, Derivatives and Law*, 1999, Kluwer Law International.

Hudson, A, *The Law on Investment Entities*, 2000, Sweet & Maxwell.

Hudson, A, *Securities Law*, 1st edn, 2008a, Sweet & Maxwell.

Hudson, A, *Understanding Equity & Trusts*, 3rd edn, 2008b, Routledge-Cavendish.

Hudson, A, *Equity & Trusts*, 6th edn, 2009a, Routledge-Cavendish.

Hudson, A, *The Law of Finance*, 2009b, Sweet & Maxwell.

Insolvency Service, *Productivity and Enterprise: Insolvency – A Second Chance*, 2001, Cm 5234 The Insolvency Service.

Ireland, P, 'Corporate governance, stakeholding, and the company: towards a less degenerate capitalism?' [1996] 23 *Journal of Law and Society* 287.

Ireland, P, 'Company law and the myth of shareholder ownership', (1999) 62 *Modern Law Review* 32.

Jeter, L, *Disconnected – Deceit and Betrayal at WorldCom*, 2003, Hoboken: Wiley and Sons, 2003.

Klein, N, *No Logo*, 2000, Flamingo.

Klein, N, *The Shock Doctrine*, 2007, Penguin.

Kollin, D and Kollin, E, *The Unincorporated Man*, 2009, Tor.

Krugman, P, *The Great Unravelling*, 2003, Penguin.

McCormack, G, 'OEICs and trusts: the changing face of English investment law', (2000) 21 *The Company Lawyer* 2.

McLean, B and Elkind, P, *The Smartest Guys in the Room*, revised edition, 2004, Penguin.

Marcuse, H, *One-Dimensional Man*, 1964, Routledge & Kegan Paul.

Masten, S, 'A legal basis for the firm', in Williamson, O and Winter, S (eds), *The Nature of the Firm*, 1993, Oxford University Press.

Mill, JS, *On Liberty*, 1859, Liberal Arts Press.

Mitchell, C, 'Lifting the corporate veil in the English courts: an empirical study', (1999) 3 *Company Financial and Insolvency Law Review* 15.

Muchlinski, P, *Multinational Enterprises and The Law*, revised edition, 1999, Blackwell.

Ottolenghi, S, 'From peeping behind the corporate veil to ignoring it completely', (1990) 53 *Modern Law Review* 338.

Parkinson, J, *Corporate Power and Corporate Responsibility*, 1993, Oxford University Press.

Rajak, H, 'The oppression of minority shareholders', (1972) 35 *Modern Law Review* 156.

Reich, R, *Supercapitalism*, 2008, Icon Books.

Roe, F, 'Political preconditions to separating ownership from corporate control', (2000) 53 *Stanford Law Review* 539.

Rubin, GR, 'Aron Salomon and his circle', in Adams, J (ed), *Essays in Honour of Clive Schmitthoff*, 1983, Professional Books.

Stiglitz, J, *The Roaring Nineties*, 2004, Penguin.

Thompson, EP, *The Making of the English Working Class*, 1963, Victor Gollancz.

Van Parijs, P, *Real Freedom for All – What if Anything Can Justify Capitalism?*, 1993, Oxford University Press.

Index